Foundations of Physical Chemistry

Charles Lawrence
Oakham School, Rutland

Alison Rodger
Department of Chemistry, University of Warwick

Richard Compton
Physical and Theoretical Chemistry Laboratory and St. John's College, University of Oxford

Series sponsor: **ZENECA**

ZENECA is a major international company active in four main areas of business: Pharmaceuticals, Agrochemicals and Seeds, Specialty Chemicals, and Biological Products.

ZENECA's skill and innovative ideas in organic chemistry and bioscience create products and services which improve the world's health, nutrition, environment, and quality of life.

ZENECA is committed to the support of education in chemistry.

OXFORD NEW YORK TOKYO
OXFORD UNIVERSITY PRESS
1996

Oxford University Press, Walton Street, Oxford OX2 6DP

Oxford   New York
Athens   Auckland   Bangkok   Bombay
Calcutta   Cape Town   Dar es Salaam   Delhi
Florence   Hong Kong   Istanbul   Karachi
Kuala Lumpur   Madras   Madrid   Melbourne
Mexico City   Nairobi   Paris   Singapore
Taipei   Tokyo   Toronto
and associated companies in
Berlin   Ibadan

Oxford is a trade mark of Oxford University Press

Published in the United States
by Oxford University Press Inc., New York

A catalogue record for this book is available from the British Library

Library of Congress Cataloging in Publication Data
(Data available)
ISBN   0 19 855904 6

Typeset by the authors
Printed in Great Britain by
Bath Press Ltd, Bath

Founding Editor's Foreword

Students embarking on chemistry degree courses at university have very disparate and fragmented knowledge bases. This Oxford Chemistry Primer is designed to bridge the gap between A level and first year undergraduate physical chemistry by providing a concise and elementary coverage of a broad range of essential topics, thus allowing all students to acquire the same basic skills required for more advanced studies.

This Primer should be of interest to all A level pupils and their teachers both at school and university. It should stimulate young people reading chemistry at advanced level at school and excite them to aspire to serve an apprenticeship in chemistry at university.

Stephen G. Davies
The Dyson Perrins Laboratory, University of Oxford

Preface

This Oxford Chemistry Primer seeks to provide a foundation in physical chemistry suitable for students starting university courses in chemistry and allied subjects. It aims to link some material familiar from school (Advanced Level) studies with a selection of essential ideas usually encountered early in the freshman year. The first two chapters consider the structure of atoms and molecules. Chapters 3, 4 and 5 discuss chemical reactivity through energetics, kinetics, and equilibria. Chapter 6 gives a brief insight into a few more advanced areas. We hope that the book will be accessible and stimulating to those studying Advanced Level chemistry whilst providing a sound basis for university work. The mathematical aspects of the subject have been minimized as far as is consistent with clarity.

We are very grateful to all those who have so generously given us much valuable constructive criticism and advice. Doubtless errors remain, of our own, for the reader to find and, we hope, relay so that they may be corrected in any future edition. In particular we thank John Freeman who created all the figures for this primer. His artistic talents and patience are much appreciated.

Oakham	C.P.L.
Warwick	A.R.
Oxford	R.G.C.

Contents

1 Atoms and ions: the building blocks of matter

1.1 Introduction

Chemistry is the science of the properties and reactions of atoms and molecules. Physical Chemistry has been defined as 'anything that is interesting'; it is usually taken to be the parts of chemistry that involve measuring or observing, then developing explanations for the experimental results, and subsequently making predictions for new situations. Physical chemists study widely different subjects, but share the same label because of the methodology they adopt. The questions physical chemists ask include:

- Why, at room temperature and pressure is ice a solid that floats, then melts, in its own liquid?
- Why is air colourless but the halogens coloured?
- Why is some rain acid?
- Why is sulfur non-conducting, silicon semi-conducting, and silver metallic?
- Why is diamond one of the hardest substances known, yet graphite an important lubricant?
- Why does fluorine react spontaneously with nearly all other elements, but helium with nothing?
- Why do some molecules cause cancer and others cure it?
- Why are some street-lights yellow, some white, some bluish, and others a pale pink?

In this book we do not attempt to answer *all* those questions, rather to provide the basis from which the answers may be understood. The focus of this chapter is atoms; in Chapter 2 we consider molecules. The subsequent chapters are devoted to the behaviour of molecules, particularly their reactions.

Because physical chemists study systems ranging in size from planetary atmospheres to atoms and measure energies ranging from those of microwaves to γ-rays emitted from radioactive nuclei, we use a wide range of prefixes on units. For example,

giga:	1 Gm	=	10^{+9}	m
mega:	1 Mm	=	10^{+6}	m
kilo:	1 km	=	10^{+3}	m
deci:	1 dm	=	10^{-1}	m
centi:	1 cm	=	10^{-2}	m
milli:	1 mm	=	10^{-3}	m
micro:	1 μm	=	10^{-6}	m
nano:	1 nm	=	10^{-9}	m
pico:	1 pm	=	10^{-12}	m
femto:	1 fm	=	10^{-15}	m

1.2 Atoms

Atoms are the building blocks of all matter. This full stop

.

measures about 0.5 mm, but the carbon atoms of which it is composed have a radius of about 0.10 nm: the full stop is 5 million atoms wide. However, although there is extensive experimental evidence for the existence of atoms, it is all indirect in the sense that we cannot *see* such small particles.

The most recent evidence for atoms uses another sense: touch. In *atomic force microscopy* (AFM) the 'finger' is a very fine tip of silicon nitride,

The first 'evidence' for the existence of atoms was the observation, by Robert Brown in 1827, of particles 'dancing' in water under a microscope. As the water and particles had been encapsulated millions of years previously inside a natural quartz cavity, the motion was not due to any living organisms.

Si₃N₄. As it is moved over the surface under study, successive atoms attract or repel the probe producing images like that of graphite in Fig. 1.1. We take as our starting point the existence of atoms.

As visible light wavelengths range from about 350 – 750 nm, the smallest particle we could possibly see, even with a light microscope, is of this order, ~ 1 μm. Atoms are ten thousand times smaller, and thus cannot be seen with light. In fact smell and taste are more sensitive than sight since we can detect very small amounts of some molecules. We can detect 1 molecule in 10^9 of 2-methyl-2-propanethiol which is added to odourless methane (natural gas) to give us what we identify in the laboratory or at home as natural gas, and only 1 molecule in 10^{11} of 2-methoxy-3-(2-methylpropyl)pyrazine in water is needed to convince us we are eating green peppers. There is current debate as to whether we can smell certain single molecules.

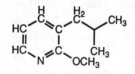

An atom is described by two quantities: Z and A. Z, the atomic number, is the number of protons in the nucleus of the atom (and also the number of electrons in the neutral atom). A, the mass number, is the number of nucleons, *i.e.* protons plus neutrons, in the nucleus. We represent an atom of element X as $^A_Z X$. If the number of electrons does not equal Z, the net charge is indicated by a superscript after the symbol. Thus an α-particle (alpha particle), a helium atom with both electrons removed, is denoted: $^4_2 He^{2+}$.

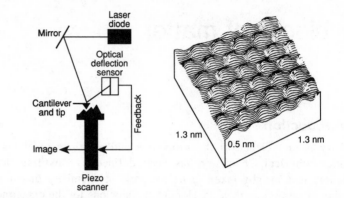

Fig. 1.1 Schematic illustration of AFM and the resulting picture of graphite. Piezoelectric sensors precisely control the tip position relative to the substrate.

1.3 Atomic structure

Since all chemistry involves atoms we need a good description of them. A rather vague picture of the atom is that of a dense nucleus, containing protons and neutrons, surrounded by a diffuse cloud of electrons. This is in fact *Rutherford's model* of the atom which followed from the first experiment that gave evidence for the existence of a dense nucleus. In 1911 a fine gold foil was bombarded with α-particles. Most of the incident α-particles passed straight through the foil, but a very small fraction of them (1 in 8000) was deflected back through significant angles. It was concluded that the foil must contain well-separated (on an atomic scale), massive particles. Since the positively charged α-particles were reflected it was further thought that the nuclei were positively charged. Atomic nuclei have in fact been found to have diameters of less than 100 fm, whereas atoms have diameters of ~ 100 pm — thus explaining why only a very small fraction of particles were deflected.

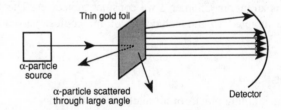

Fig. 1.2 Rutherford scattering of alpha particles.

There are two possible problems with the Rutherford model of the atom.
1.   If the electrons are stationary, they will be attracted into the nucleus.
2.   If, on the other hand, the electrons revolve around the nucleus then they are accelerating, so should radiate energy and spiral into the nucleus.

In both cases, the atom is not stable.

Bohr, in 1913, neatly side-stepped these problems simply by postulating that there are certain stable orbits in which an electron can revolve without radiating energy. These orbits are those with electron angular momentum of $n\hbar = nh/(2\pi)$, for $n = 1, 2, \dots$. He also proposed that each orbit has a particular energy:

$$E = \frac{-m_e e^4}{8\varepsilon_o^2 h^2 n^2} = -R_H \left(\frac{1}{n^2}\right) \tag{1.1}$$

where $R_H = 2.179 \times 10^{-18}$ J is known as the Rydberg constant for hydrogen, $h = 6.6256 \times 10^{-34}$ Js is Planck's constant, $\varepsilon_o = 8.854 \times 10^{-12}$ J^{-1}C^{2}m^{-1} is the vacuum permittivity, $m_e = 9.109 \times 10^{-31}$ kg is the mass of an electron, and $e = 1.602 \times 10^{-19}$ C is its charge. A consistent set of units for eqn (1.1) is

$$[\text{joules}] = \frac{[\text{kg}]\,[\text{C}]^4}{\left\{[\text{C}]^2[\text{J}]^{-1}[\text{m}]^{-1}[\text{J}][\text{s}]\right\}^2}$$

Fig. 1.3 shows the allowed energy levels for an electron in a hydrogen atom.

The energy levels of hydrogen cannot be measured directly. All that can be measured are the *well-defined* amounts of energy required to cause the electrons to *jump* between orbits or levels. Jumps between energy levels can be made either by absorbing (increasing) energy or emitting (decreasing) energy. We usually measure atomic emission spectra, in which case, n_2, the initial level, is larger than n_1, the final level. The energy emitted is given by the Rydberg equation:

$$\Delta E(n_2 \rightarrow n_1) = E(n_2) - E(n_1) = R_H \left(\frac{1}{n_1^2} - \frac{1}{n_2^2}\right) \tag{1.2}$$

A particle that is constrained to move in a circle, such as a ball on a string, is constantly accelerating because its linear velocity is constantly changing. The *angular momentum* of the ball in the figure below is mvr, where m is its mass, v its linear velocity, and r the radius of the circle around which it moves. The lowest energy Bohr orbit has a radius of $a_o = 5.292 \times 10^{-11}$ m.

The Rydberg formula for the positions of lines in the hydrogen spectrum was worked out long before Bohr's postulate led to any understanding of what was being observed. A number of series with final state n_1 as in the table below and $n_2 > n_1$ have been identified. Ultraviolet light is emitted in the Lyman series, visible light in the Balmer series, and infra red radiation by the others.

n_1	series
1	Lyman
2	Balmer
3	Paschen
4	Brackett
5	Pfund

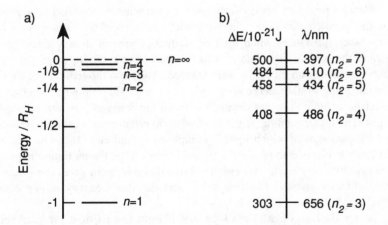

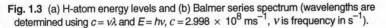

Fig. 1.3 (a) H-atom energy levels and (b) Balmer series spectrum (wavelengths are determined using $c = v\lambda$ and $E = hv$, $c = 2.998 \times 10^8$ ms^{-1}, v is frequency in s^{-1}).

Despite its successes, it quickly became apparent that the Bohr model of the atom could not be the whole story as it could not account for the spectra of atoms other than H, nor could it be extended to explain why some of the

What we define to be the zero point of energy is arbitrary. For hydrogen we choose zero to be the point where the electron and proton are infinitely far apart, *i.e.* $n_2 = \infty$. All other levels are more stable than this one, so have negative energies.

lines in the hydrogen atom spectrum were in fact two or more lines very close together in energy.

1.4 Electrons in atoms: the development of quantum mechanics

The resolution of the problems inherent in the Bohr model of the atom was provided by *quantum mechanics* which was developed in the 1920s. Quantum mechanics is the name given to the theory that starts with the postulate that energy comes in well defined 'quanta' or packets and then uses classical mechanics as much as possible. In some ways Bohr had already made this leap when he suggested that hydrogen has well-defined orbits of well-defined energy, but it needed Planck to recognize that light also comes in *discrete quanta* whose energy, E, is dependent upon the frequency, v, of light according to

$$E = hv \tag{1.3}$$

The big conceptual leap that is made in accepting Planck's relationship is that we acknowledge that light energy is in discrete packages, called photons, not a continuum.

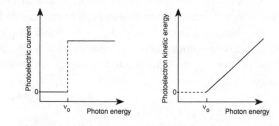

Fig. 1.4 Schematic representation of the photoelectric effect.

Planck's postulate immediately resolved the dilemma that had been posed by the 'photoelectric effect', in which electrons are ejected from a metal (or any other substance) when light of frequency greater than a given value strikes the surface. By analogy with the macroscopic world, we might have expected that if no electrons were emitted as soon as the light shone on the sample some would emerge after a period of time. This is not so. Whether an electron is emitted or not depends only on the *frequency* of the incoming light. The understanding of the photoelectric effect comes through *photons* — the packages of which light is composed — and eqn (1.3). One photon causes one electron to be ejected if, and only if, it has the minimum required energy, hv_0; any additional energy above this minimum gives the electron extra kinetic energy. Photons below the threshold energy *never* cause electrons to be ejected (Fig. 1.4).

It was de Broglie who realized that if light has particle (or 'package') character, as proposed by Planck, as well as wave character, then other particles such as electrons may also have both particle and wave properties. This is referred to as wave–particle duality. De Broglie proposed that the relationship between momentum, p, (and hence mass, m, and velocity, v) and wavelength, λ, is

---

The concept of energy is central to much of chemistry. It is very important to understand what an energy value that we measure or calculate means. Two approaches are as follows.

• Consider units to see how the energy relates to other properties such as force, *e.g.*

$J = N \times m$,

Energy = Force × distance

• Relate a new situation to one with which you are already familiar. For example, consider 50 kJ of energy and think what happens when you heat 100 cm^3 of 25°C water in a 500 W = 500 J s^{-1} microwave oven for 100 s. The oven will produce 50 kJ of energy. If the water only just reaches 100°C then, since the heat capacity of water is about 4 J / g / °C, the water has absorbed 30 kJ of energy and you conclude that the heating process is 60% efficient.

Problem: Green light has frequency 6×10^{14} s^{-1}. What is the energy of one such photon?

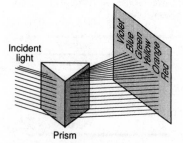

Wave–particle duality is difficult to rationalize, but the fact is that light and electrons (and all other particles) behave as waves if we look for wave properties such as dispersion and behave as particles if we look for particle properties such as momentum transfer. Dispersion of white light by a prism is illustrated above.

$$p = mv = \frac{h}{\lambda} \tag{1.4}$$

Schrödinger's contribution to the development of our detailed understanding of the chemical world was showing how to write the mathematical equation for the *wavefunction*, the function that describes how the particles (electrons, protons, *etc.*) in the system behave. Writing the Schrödinger equation for an atom, molecule, or other system requires us first to write down the equation for its classical kinetic energy (*KE*) and potential energy (*PE*). Then it is possible to write down the Schrödinger equation. To solve the equation we also need to know the so-called *boundary conditions*, which tell us where the wavefunction that describes the particle-wave is zero.

The simplest system we could examine is an electron in a box with infinitely high sides and zero potential energy inside the box. The boundary conditions require that the particle cannot get out of the box, so the wavefunction must be zero outside. *KE* is simply

$$KE = \tfrac{1}{2} m_e v^2 = \frac{p^2}{2m_e} \tag{1.5}$$

Fig. 1.5 is a schematic illustration of how to write and then solve the Schrödinger equation for the electron in a box. As the hydrogen energy level diagram (Fig. 1.3) might lead us to expect, we get an infinite number of discrete solutions to the Schrödinger equation for this system.

If a stationary electron is accelerated through a 1 kV = 1000 J/C potential difference, it will acquire 1000 e J of kinetic energy, so its momentum is
$$p = \sqrt{2 \times 1000 m_e e}$$
It follows that its wavelength is 39 pm, which compares with a typical chemical bond length of 100 – 200 pm.

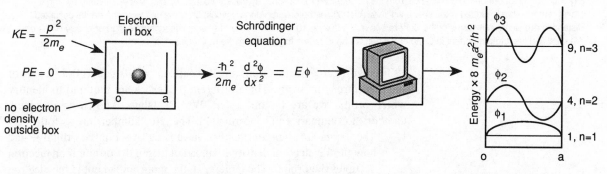

Fig. 1.5 The 'Schrödinger machine': the Schrödinger equation and its solutions for an electron in a box with infinitely high walls. Wavefunctions are usually written using Greek letters such as ϕ, (phi), rather than Roman letters such as f.

1.5 The hydrogen atom

The Schrödinger equation for the hydrogen atom is not quite as simple as that for the electron in a box as there is the potential energy due to the attraction of the electron and proton, so we must add to the Schrödinger machine

$$PE = \frac{-e^2}{4\pi\varepsilon_0} \times \frac{1}{r} \tag{1.6}$$

where r is the distance between the proton and the electron. The solutions to the Schrödinger equation are the wavefunctions for hydrogen (Fig. 1.6).

It should be noted that in practice we can only solve the Schrödinger equation analytically for very few systems including the particle in a box and the hydrogen atom.

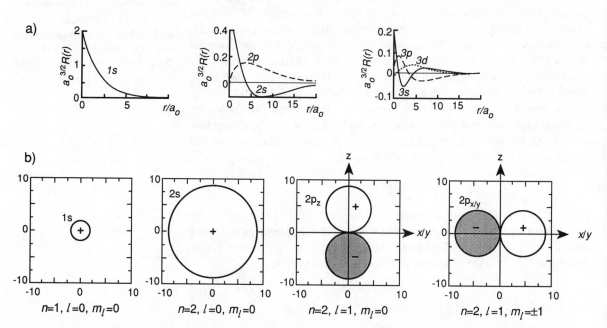

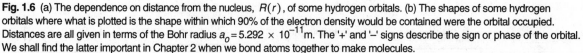

Fig. 1.6 (a) The dependence on distance from the nucleus, $R(r)$, of some hydrogen orbitals. (b) The shapes of some hydrogen orbitals where what is plotted is the shape within which 90% of the electron density would be contained were the orbital occupied. Distances are all given in terms of the Bohr radius $a_o = 5.292 \times 10^{-11}$m. The '+' and '−' signs describe the sign or phase of the orbital. We shall find the latter important in Chapter 2 when we bond atoms together to make molecules.

The hydrogen wavefunctions are also called *orbitals* due to Bohr's idea of single electrons in *orbits*. Three different properties are required to identify which orbital we are talking about. We use labels, known as *quantum numbers*, to summarize this information. The three numbers are as follows.

1.   The *principal quantum number*, n, tells us how big the orbital is and how far the electron density is spread out from the proton if an electron occupies the orbital. The larger n is, the more spread out is the electron density and the smaller is its attraction to the proton. Thus we would expect larger values of n to correlate with smaller energies. This is precisely what eqn (1.1) says. The energy of an electron in a *hydrogen* orbital is *completely* determined by its principal quantum number.

2.   The *orbital quantum number*, l, describes the shapes the electron density adopts. If $l = 0$, then the electron density is spherical (so-called *s* orbitals); if $l = 1$ then it has the dumbbell shape shown in Fig. 1.6 (so-called *p* orbitals). Higher values of l have increasing numbers of lobes. For a given n, l can take values from 0 to $n-1$.

3.   The *angular momentum (or azimuthal) quantum number*, m_l, describes which direction the orbital is pointing. m_l takes integer values from $-l$ to l. The functions with well-defined values of m_l are often complex (in

The apparently arbitrary choice of letters to label the shapes of the hydrogen orbitals comes from the appearance of lines in the hydrogen spectrum. They were labelled sharp $l = 0$, principal $l = 1$, diffuse $l = 2$, and fundamental $l = 3$.

the mathematical sense) and so we usually take combinations of them to give the real functions shown in Fig. 1.6. This results in orbitals such a p_x which is a p orbital (with $l = 1$ and m_l not well defined) pointing along the x-axis.

1.6 Hydrogen-like systems

Hydrogen-like systems are those such as Li^{2+} that have only one electron, but a higher nuclear charge and nuclear mass than hydrogen. Eqn (1.6) then needs to be modified to account for the greater attraction of the electron for the more highly charged nucleus; a factor of Z is all that is required. The orbital energies follow from eqn (1.1) by replacing the Rydberg constant for hydrogen, R_H, by R_X for atom X, where m_X is the atomic mass of X

$$R_X = Z^2 R_H \left(1 + \frac{m_e}{m_H}\right) \Big/ \left(1 + \frac{m_e}{m_X}\right) \cong Z^2 R_H \qquad (1.7)$$

Thus, for a given n, the orbitals are energetically more stable for $Z > 1$ nuclei than for hydrogen and also spatially contracted towards the nucleus.

1.7 Many-electron atoms

As soon as we have more than one electron we can no longer solve the Schrödinger equation because the effects of electron-electron repulsion in the potential energy term make the equation too complicated. It turns out, however, that the behaviour of each electron in a many-electron system can, to a first approximation, be described by an orbital similar to a hydrogenic one. Thus we assign each electron to an orbital that is labelled with the same set of quantum numbers as used for hydrogen.

One difference between the orbitals of a hydrogenic system and those of many-electron systems is that the energy of an electron in an orbital is not determined by n alone. The energy level diagram for lithium, illustrated in the margin, shows this. We can see why ns and np electrons have different energies if we consider the diagrams in Fig. 1.6a. A $2s$ electron is on average closer to the nucleus than a $2p$ electron would be. This means the $2p$ electron is more *shielded* from the nucleus by the $1s$ electrons than is the $2s$ electron which *penetrates* closer towards the nucleus than does the $2p$ electron. The $2p$ electron therefore feels less attraction towards the nucleus than does a $2s$ electron, so the *degeneracy* (having the same energy) of the $2s$ and $2p$ orbitals for hydrogen is lost in many-electron systems. The three $2p$ orbitals are still degenerate because they point in different directions and do not shield each other.

A further factor that must be considered for many-electron atoms was discovered in the famous Stern–Gerlach experiment where a beam of silver atoms, which have an odd number of electrons, was fired through an inhomogeneous magnetic field. The incident beam of atoms was split into two equal halves. The only explanation for this result was that electrons have one of two spins; thus a fourth quantum number: m_s which can take values of $+\frac{1}{2}$ or $-\frac{1}{2}$ was introduced for each electron. The four quantum numbers needed to identify an electron in an orbital are summarized in Table 1.1.

The electron density in an occupied orbital, ϕ, is $|\phi|^2$. So the electron density in an occupied hydrogen orbital is described by the square of the orbitals in Fig. 1.6.

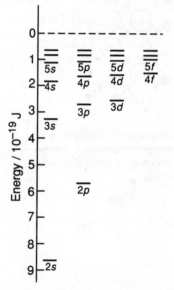

Orbital energy level diagram for lithium.

The constituent particles of atoms, electrons, protons, and neutrons all have spins of $+\frac{1}{2}$ or $-\frac{1}{2}$. Such particles are called *fermions*.

Table 1.1 Quantum numbers.

name and symbol	possible values	orbital information
principal, n	$1, 2, ...$	energy and size
orbital, l	$0, 1, ..., n-1$	shape
angular momentum, m_l	$-l, -l+1, ..., l-1, l$	orientation
spin, m_s	$-\frac{1}{2}, \frac{1}{2}$	spin

Shells, sub-shells, and orbitals

Chemists frequently use the label *shell* to refer to all the orbitals with a given principal quantum number, n. The label *sub-shell* refers to the l quantum number, so there are n sub-shells in the nth shell (Table 1.1). Thus the first shell is the $1s$ sub-shell, the second shell contains the $2s$ and $2p$ sub-shells, the third shell contains the $3s$, $3p$, and $3d$ sub-shells, and the fourth shell contains the $4s$, $4p$, $4d$, and $4f$ sub-shells. A close inspection of the lithium orbital energy level diagram in the margin above shows that the energy ordering is not simply $1s$, $2s$, $2p$, $3s$, $3p$, $3d$, $4s$ ... . The usual orbital energy ordering is given in the margin in a diagrammatic form. To a first approximation electrons are assigned to sub-shells in increasing order of energy. However, as n increases the difference in energy between sub-shells decreases and other factors may be more important.

The following three rules may be used to assign electrons to orbitals. Half arrows are used to indicate electrons, with the direction denoting spin up $(+\frac{1}{2})$ or down $(-\frac{1}{2})$.

(1) Only two electrons of opposite spins can be assigned to an orbital. A pair of electrons in an orbital is described as *spin-paired*.

(2) Electrons are assigned to orbitals one at a time with the next electron being assigned to the lowest energy orbital available (not already occupied by two electrons).

(3) If the lowest energy available orbitals are degenerate (have the same energy) then one electron is first assigned to each orbital (beginning with the highest m_l orbital) with the spin of all the single electrons being the same. Then a second electron of opposite spin is assigned to each orbital in turn.

The above rules form the *aufbau* or building-up principle. They ensure that no two electrons in an atom have the same four quantum numbers; this is a version of the *Pauli exclusion principle*.

One additional factor that must be considered when we try to decide how to allocate electrons to orbitals to determine the lowest energy state of an atom is the experimental result that half-filled and filled sub-shells are particularly stable. If the lowest available orbital already contains one electron and the next lowest orbital is *very* close in energy then, as for copper, whose electron configuration is $(1s)^2(2s)^2(2p)^6(3s)^2(3p)^6(4s)^1(3d)^{10}$, the orbital energy ordering may be over-ruled.

One indication of energy ordering of orbitals is given by *ionization energies* (E_i, the energy required to remove an electron from an atom in the gaseous phase) and *electron affinities* (E_{ea}, the energy released when an

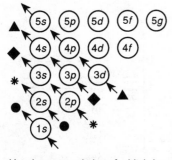

Usual energy ordering of orbitals in a many-electron atom. Follow the arrows to go from lower to higher energy.

Electronegativity is a label describing both how tightly an atom holds on to its electrons and how much it attracts additional electrons. One definition (that due to Pauling) is that electronegativity equals the average of the ionization energy and the electron affinity. Elements on the right-hand side of the periodic table are most electronegative, while those on the left-hand side are least electronegative.

electron is added to an atom in the gaseous phase). These are plotted for first and second row atoms in Fig. 1.7.

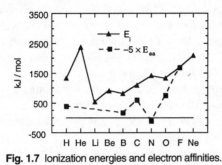

Fig. 1.7 Ionization energies and electron affinities.

There is a general increase in ionization energy across a period. The simple linear trend for the second row is disrupted at Li, B, and O implying either that He, Be and N are particularly stable and/or Li, B and O are particularly unstable. N also has a very low electron affinity. These observations may be understood as follows.

- The change in ionization energy between He and Li is due to the fact that the electron to be ionized from Li is in a $2s$ orbital whereas that from He is in a $1s$ orbital. Being so much closer to the nucleus the helium electron is held much more tightly.
- The Be/B step correlates with the point where the last electron to be added was $2s$ and the next is $2p$ and we know that $2p$ electrons are attracted less strongly to the nucleus than are $2s$ electrons, which penetrate closer in and also shield the $2p$ electrons from the nucleus.
- At the N/O step, N has a half-filled sub-shell and O has two electrons spin-paired in a $2p$ orbital and two half-filled $2p$ orbitals. The double occupancy of an orbital causes some degree of repulsion between the electrons which is unfavourable; in addition, half-filled (and filled) sub-shells are also special in that they have a maximum amount of what is known as *exchange energy*, which is a quantum mechanical contribution to the stability of a system from electrons with the same spin in degenerate orbitals. The electron configuration of Cu which has fully occupied $3d$ orbitals and only one $4s$ electron is one consequence of the special stability of filled sub-shells.

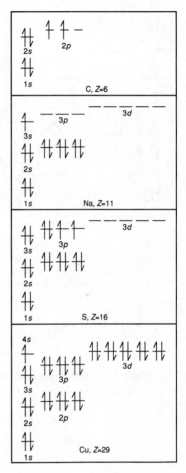

1.8 The periodic table

Originally the periodic table came from putting the atoms in ascending mass order and lining up in columns the atoms with similar chemical properties. Since chemical properties are governed largely by where the outermost electrons are and how tightly they are held by the nucleus, the same table follows from considering quantum numbers. By lining up in rows those atoms with the same maximum n in their occupied orbitals, and in columns those atoms whose highest occupied orbitals have the same value of l and m_l, the periodic table of Fig. 1.8 results. More usually we extract the lanthanides and actinides and put them at the bottom as in the periodic table

The periodic table is best thought of in terms of ascending atomic number, Z, (number of protons), rather than ascending mass (protons plus average number of neutrons) because protons and electrons determine chemistry. Ar has mass 40.0 units and $Z = 18$, but K has mass 39.1 units and $Z = 19$.

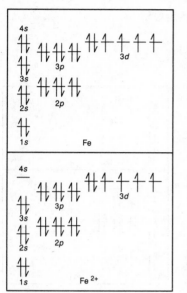

inside the back cover. The columns are numbered from 1 to 18 from left to right of the standard table.

We use the term *valence* orbitals (and electrons) to refer to the orbitals that are newly occupied across a row of the periodic table. We also define 'blocks' of the periodic table in terms of the highest energy occupied sub-shell. The *s*-block contains the alkali (Group 1) and alkaline earth (Group 2) metals; the *p*-block contains Groups 13–18; the *d*-block contains the transition metals and makes up Groups 3–12; and the *f*-block contains the lanthanides and actinides. Some points to note about the *d*-block include:

- The *d*-block sits between the *s*- and *p*-blocks because of the energy ordering of the orbitals.
- The principal quantum number for a *d* orbital in a given row of the periodic table is one less than the *s* and *p* orbitals in the same row.
- The easiest electron to remove from an atom in the *d*-block is an *s* electron. (There is no simple explanation for this.) The electron configuration for Fe and Fe^{2+} are illustrated in the margin.

Core electrons are those that are 'buried' within the *valence shell* of electrons. The electron configuration for the core orbitals may be summarized by the notation [X], where X is the noble gas (Group 18) of the preceding row of the periodic table. This notation is used in the periodic table inside the back cover.

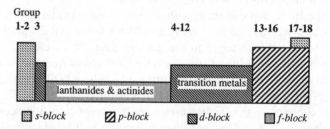

Fig. 1.8 Schematic periodic table indicating partially filled valence orbitals and the positions of the lanthanides and actinides.

If chemistry were to be limited to only one topic, the periodic table, then we would have hardly limited the subject at all since the periodic table summarizes so much information if we but know how to interpret it. For example, it tells us the sizes of atoms since the size of the valence orbitals increases down the table as *n* increases and decreases across the periodic table as *n* remains constant but the nuclear charge increases and attracts the electrons. Thus it tells us how easy it is to remove electrons from the atom; and it tells us a great deal about the chemistry of the atoms since chemistry is dominated by the valence electrons.

1.9 Atomic ions

We have looked at ionization energies (the removal of electrons and the production of positive ions) and electron affinities (the addition of electrons and the production of negative ions) in the context of the stability of shells and half shells. For many-electron atoms, more than one electron may be removed or added, resulting in ions such as Al^{3+} and O^{2-}. It is generally true that elements on the left-hand side (*s*- and *d*-blocks) of the periodic table form positive ions, known as *cations*, and those on the right-hand side of the periodic table form negative ions, known as *anions*. Elements in the middle are less likely to be found as ions; instead they share their electrons and form what are known as *covalent bonds* as we shall see in Chapter 2.

1.10 Aggregates of atoms: the states of matter

So far in this chapter we have been considering atoms (and briefly ions) in isolation and looking at their atomic structure. In practice we rarely find isolated atoms. To discuss the properties of matter we must consider their possible states. Happily most substances exist obviously as either solids, liquids, or gases as illustrated in Fig. 1.9. This state depends on the strength of the attractive forces between the atoms (or molecules) compared with the kinetic energy they have at a given temperature.

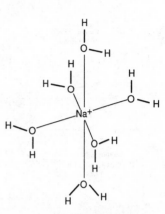

Some anions are more stable than their neutral atoms (Fig. 1.7). However, it *always* takes energy to form a cation, so it is nonsense to say 'Na^+ is more stable than Na'. Na^+ is only found naturally when there is a compensating stabilization provided by solvation, as shown in the hydrated Na^+ ion above, or by ionic bond formation.

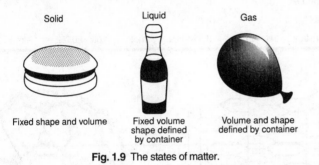

Solid	Liquid	Gas
Fixed shape and volume	Fixed volume shape defined by container	Volume and shape defined by container

Fig. 1.9 The states of matter.

Consider the three elements carbon, mercury, and helium. Carbon exists as a solid, denoted (s), in various allotropes (Fig. 1.10) at SATP (standard ambient temperature and pressure). The forces between carbon atoms are very strong and are discussed in Chapter 2. Even to liquefy carbon requires high temperatures (3730°C for graphite). Mercury is one of two elements in the periodic table that is liquid, (ℓ) at SATP (bromine is a molecular, not an atomic liquid, at SATP). Finally, helium, the least reactive element, is a gas, (g), at SATP; the forces between helium atoms are so small that its boiling point is 4.2 K at atmospheric pressure. These three elements demonstrate the extremes of stability and structure in our material world.

The gaseous state

Early experiments on gases resulted in three empirical laws (that apply to both atoms and molecules) relating the different quantities required to define the state of a gas (Table 1.2).

- If the temperature, T, and the pressure, P, are held constant, then, as the amount of gas, n moles, is increased, the volume, V, increases according to

  Avogadro's Law: V/n = constant (for fixed T and P).

- If n and T are held constant, then the reduction in V caused by increasing P is given by

It is not always obvious whether some systems are solid, liquid or gas.
- Glass looks like a solid, but is in fact a liquid. Careful inspection of old bottles or stained glass windows will show the glass is thicker at the bottom than at the top.
- The molecules of a liquid crystal are ordered only in one or two dimensions and the resulting system behaves as a cross between a liquid and a solid.
- Superfluids, such as liquid He-II, are liquids but have no viscosity.

Empirical means 'from experiment'.

A warm room has a temperature of 25.0°C, or 298.15 K; atmospheric pressure is usually about 1.013×10^5 Pa or 1 atm. These values are accepted as the SATP. Standard temperature and pressure (STP) is 273.15 K (0°C) and 1.013×10^5 Pa.

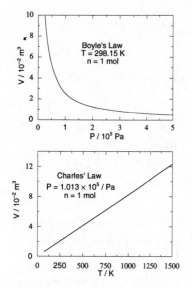

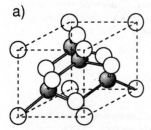

Boyle's Law: PV = constant (for fixed n and T).

- If n and P are held constant, then as T increases V also increases according to

Charles' Law: V/T = constant (for fixed n and P) with T in K.

We can combine these three laws to give the

Ideal Gas Law:

$$PV = nRT \qquad (1.8)$$

where R is known as the universal gas constant and T is in K.

Table 1.2 The four quantities required to define the state of a gas. 1 mole (1 mol) is the amount of substance containing 6.0223×10^{23} particles (atoms, molecules, ions *etc.*).

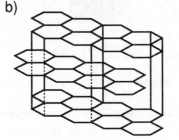

Quantity	Symbol	Unit (SI)	Alternative unit
pressure	P	$Pa = N\,m^{-2}$	atmospheres, atm
volume	V	m^3	dm^3
temperature	T	K	
amount	n	mol	
gas constant	R	$8.31434\ J\,K^{-1}\,mol^{-1}$	$0.0820562\ dm^3\,atm\,K^{-1}\,mol^{-1}$

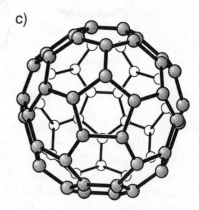

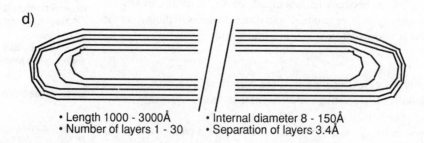

- Length 1000 - 3000Å
- Number of layers 1 - 30
- Internal diameter 8 - 150Å
- Separation of layers 3.4Å

Fig. 1.10 Allotropes of carbon: (a) diamond (some of the carbons are shaded for clarity), (b) graphite, (c) the fullerene C_{60} (Bucky ball), and (d) a schematic drawing of a Bucky tube. A Bucky tube consists of a layer of graphite, 'rolled up' into a tube and 'capped' with a unit made of C_5 and C_6 units. Electron microscopy shows that a typical tube can contain between 1 and 30 tubes inside each other, each layer separated by 340 pm (as in graphite). (See M.L.H. Green *et al.* Nature *372* (1994), 159 for further details.) A single shell of a Bucky tube is illustrated on the front cover.

The ideal gas law is one of the most important in physical chemistry. It enables us to calculate the value of any one of P, V, n, or T, provided values for the other three are known. As with any empirically derived law, we must ask what are its limitations? Two assumptions underlie the ideal gas law:
(1)   atoms (and molecules) are points,
(2)   atoms (and molecules) do not interact with one another.
We know that (1) is not strictly true as we have already discussed the size of atoms. However, they are very small so it is almost true unless we are working at very high pressures. (2) is also not true — otherwise we would not observe liquids and solids. However, it is surprisingly close to the truth for most purposes, even when we are considering the vapour above a liquid.

The liquid state

The liquid state is most simply seen as intermediate between gases and solids. There are forces between the atoms (or molecules) that lead to varying degrees of structure, but the atoms are not located in well-defined positions and they readily move from place to place. Less is known about liquids than about gases and solids because of this intermediate behaviour.

The solid state

Unlike gases and liquids, solids are (generally) well ordered. This means we can probe their structure using techniques such as X-ray diffraction which depend upon the existence of 'infinite' regular lattices of points.

a)

b)

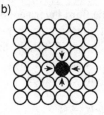

c)   d)

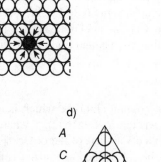

e)

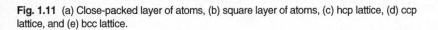
Fig. 1.11 (a) Close-packed layer of atoms, (b) square layer of atoms, (c) hcp lattice, (d) ccp lattice, and (e) bcc lattice.

Problem: Show that the volume of 1 mole of gas at SATP is 24.5 dm^3, and at STP is 22.4 dm^3. Determine the volume per atom under these two conditions.
Warning: Make sure that the units of R are consistent with the P and V units in any calculation (see Table 1.2).

Interatomic (between atoms) forces in an *atomic* liquid are quite weak, always present, and always attractive. The atoms are kept at a minimum distance from each due to electron cloud repulsion.

To understand these close-packed structures, and the relationships between the close-packed layers, build up the layers with little balls of plasticine or blue-tack.

Allotropes are different physical forms of the same element.

Problem: Determine what percentage of the volume is occupied in close packed (ccp or hcp) arrangements of spheres. How much less is occupied in a bcc arrangement?
Answer: 74% is occupied for ccp and hcp. 68% is occupied in bcc.

Note: We number the groups of the periodic table as in Fig. 1.8. An alternative numbering is from I to VIII across the *s* and *p* blocks. Thus I ≡ 1, II ≡ 2, III ≡ 13, IV ≡ 14, *etc.*

Note the similarity between diamond (Fig. 1.10) and zinc blende (Fig. 1.12).

We can classify solids into four types:
- *simple atomic or molecular* such as helium and other noble gases, and the halogens
- *metallic* such as platinum
- *ionic* such as NaCl, CsCl or $MgCl_2$
- *covalent* such as carbon in its allotropes (Fig. 1.10).

We shall look at some of the simplest cases.

Simple solids and metals: the packing of spheres

The simplest lattices are the ones that follow from the close-packing of identical spheres. In two dimensions, six spheres may be close-packed around a single sphere (Fig. 1.11a). Call this layer *A*. If atoms are placed in a second layer, *B*, above *A*, using alternate depressions between the atoms in *A*, we have the beginnings of a *three-dimensional close-packed structure*. When we consider adding a third layer in the depressions between the *B* atoms we have a choice. If the atoms of the third layer line up with those of *A*, we have *ABA* packing. This arrangement is hexagonal (Fig. 1.11c), so is called hexagonal close-packed, hcp.

If, however, we assemble the first two layers as for hcp, but place the third layer so it has its atoms in the depressions made by *B* that *do not* align with the *A* atoms, we have a different layer, denoted *C*. This packing is referred to as *ABC* and is described as cubic-close packed, ccp, since a cube can be identified if the system is tilted (Fig. 1.11d).

Another lattice is body-centred cubic (bcc) where layer *A* is a square array of atoms (Fig. 1.11b), the atoms of layer *B* sit in every depression of layer *A*, and the third layer necessarily lines up with *A*.

Most metals adopt hcp structures; some (Co, La, Pr, and Nd) are either hcp or ccp; some (Ca, Sr, Sc, Pb, Ce, Tm and Groups 9, 10, 11) are ccp; and Group 1, 5, and 6 metals as well as Ba, Fe, and Eu are bcc with eight atoms fitted around each atom.

Ionic solids

An example of an ionic solid is sodium chloride, which is usually written NaCl, but is more accurately written as $(Na^+Cl^-)_\infty$, where the 'infinity' subscript is used to indicate that a crystal has of the order of 10^{23} ion pairs. The reason for the atoms being best described as ionized is that the energy required to ionize Na is less than the energy released when Cl accepts an electron and the ions are electrostatically attracted to each other.

In many ionic solids the anions are larger than the cations, so we can view an ionic solid as a lattice of anions with cations in holes between anions. Consider either the hcp or ccp lattice discussed above. When the *B* layer atoms are placed in the depression created by three of the *A* atoms, a hole surrounded by four spheres (one above and three below) is created; we refer to such a site as a tetrahedral site.

When ZnS forms, the S^{2-} ions adopt either a ccp or an hcp lattice, and the Zn^{2+} ions occupy the tetrahedral sites. The ccp lattice makes zinc blende (Fig. 1.12) and the hcp lattice makes wurtzite (Fig. 1.12).

Now imagine the site above three spheres of *A* that is *not* occupied in *B*. In a ccp lattice (Fig. 1.11d) there are three spheres around this site in the *B*-layer that are skew with the three *A* spheres. This creates an octahedral site. If the spheres are Cl^- ions, and the Na^+ ions are put into the octahedral sites, then the sodium chloride lattice illustrated in Fig. 1.12 results.

CsCl adopts a simple cubic (not close-packed) lattice of Cl^- ions with the Cs^+ atoms in the centres of each Cl^- cube. CaF_2 is essentially the same lattice with alternate cations removed.

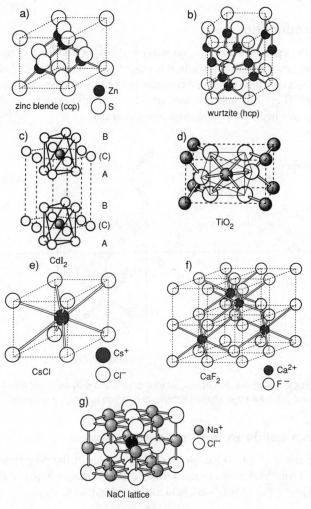

Fig. 1.12 Some crystals.

Covalent solids

Diamond (Fig. 1.10) is an example of a pure covalent solid. Such solids are very stable, being held together by strong atom–atom covalent bonds. We shall look at them in more detail in the next chapter where we shall use covalent solids as the stepping stone from atoms to molecules.

2 Molecules: the beginning of chemistry

2.1 Introduction

Chemistry begins when atoms and ions are joined or bonded together to form molecules and molecular ions. In this chapter we focus on what holds atoms together to make molecules and consider the variety of (non-spherical) shapes that result (Fig. 2.1) and the reasons why some molecular geometries are favoured. We begin by considering atoms in solids.

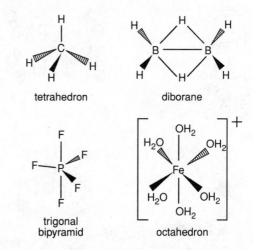

tetrahedron

diborane

trigonal bipyramid

octahedron

Fig. 2.1 Some molecules and molecular ions of different shapes. Solid triangles denote bonds coming out of the page, shaded triangles denote bonds going into the page.

2.2 From solids to bonds between atoms

The discussion of §1.10 has already taken us part of the way from isolated atoms and ions to describing the various states of matter in particular solids. The different types of solids are held together in different ways.

(1) *Simple atomic solids* such as argon form when the temperature is low enough for the weak attractive forces (known as van der Waals forces) which exist between all atoms to overcome the kinetic energy of the atoms. The atoms retain their separate identity.

(2) *Metallic solids* are held together by so-called *metallic bonds*. They result from the valence electrons of the metals losing their association

with a single atom and becoming a 'sea' of electrons which are attracted to all the neighbouring atoms. This so-called 'delocalized sea of electrons' is responsible for most of the distinctive properties of metals. This type of bonding is not found in molecules, though the delocalized metal–metal bonding in molecules such as $Rh_4(CO)_{12}$ (Fig. 2.2) has features in common with it.

(3)   *Ionic solids* such as sodium chloride are held together by electrostatic attractions between positive cations and negative anions. The attractions between the ions may be described as an *ionic bond*. If an ionic solid were to be vaporized, we would find units such as $NaCl(g)$ or $CaF_2(g)$ held togther by ionic bonds whose strengths
- depend on the relative charges of the ions,
- depend inversely on the distance between them,
- involve atoms from the left-hand side of the periodic table (which donate electrons) and the right-hand side (which accept electrons).

Ionization energies, E_i, decrease down any group in the periodic table. As one works down a group both the atomic number, Z, and the distance between the nucleus and valence electrons increase. The increase of Z would cause E_i to increase. However, the decrease in attraction of the electron by the nucleus due to its greater distance from the nucleus *and* the fact that core electrons shield the electron from the increased nuclear charge result in a net decrease of E_i down a group. The same is true for electron affinities.

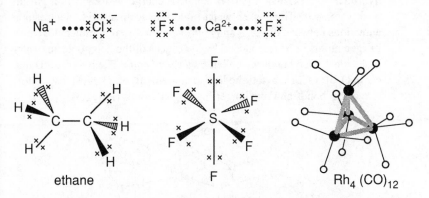

Fig. 2.2 Ionic bonds (dotted lines), covalent bonds (solid lines), and delocalized bonds (fuzzy lines) in some gaseous molecules. Crosses denote valence electrons, open circles denote CO ligands, and solid circles denote rhodium atoms.

(4)   *Covalent solids* include diamond and graphite (Fig. 1.10), silicon carbide, and silicon dioxide. Diamond is composed solely of carbon atoms that all share their outer electrons equally, with every carbon being surrounded by four carbons equivalent to it. The shared electron density can be visualized as being stretched out along the line connecting neighbouring atoms in the solid. The high electron density lines are referred to as *covalent bonds*.
- A covalent bond is formed between atoms of similar electronegativity where neither atom may be described as donating electrons to the other; rather they share their valence electrons.
- The strength of a covalent bond between two atoms is influenced by both the relative positions of the two bonding atoms and also

Each atom in graphite covalently bonds to three atoms in the same layer (Fig. 1.10). The layers are held together by weak van der Waals forces – hence the lubricant properties of graphite.

Problem: Diamond is less stable than graphite. It requires high temperature and pressure to form diamonds. Why then do diamonds not change to graphite under ambient conditions? (Hint: see Chapter 4)

18 *Molecules: the beginning of chemistry*

by the other bonds each atom makes (since the electrons available for bonding will be affected by this).

Silicon carbide resembles diamond with each C atom having four Si atoms tetrahedrally arranged about it (and vice versa). It thus adopts either the zinc blende or wurtzite structures of Fig. 1.12 — though the bonds between the atoms in silicon carbide are covalent rather than ionic. In contrast to diamond, there is not exactly equal sharing of the electron density in silicon carbide as carbon is more electronegative than Si, so each C gains somewhat more electron density than it gives.

Problem: The regular array of water molecules in ice is lost when it melts. How does this explain the fact that water at atmospheric pressure is most dense at 4°C, whereas it is only solid below 0°C?

(5)   There is a final type of solid, *the molecular solid*, which is composed of atoms covalently bound into molecules which are themselves held together by attractive forces that are weaker than the *intra-molecular* (within molecule) covalent bonds.

One example of *inter-molecular* (between molecules) forces is found in solid water. Ice is composed of water molecules whose O–H bonds are covalent, but since oxygen is much more electronegative than hydrogen each O has a partial negative charge and each H a partial positive charge. Thus, when water molecules form a solid, electrostatic attractions between oxygens and hydrogens on different molecules cause oxygen atoms to be surrounded by hydrogens of the neighbouring water molecules. Such bonds are called *hydrogen bonds*. The net result is that each O atom has a tetrahedral arrangement of H with two short (covalent) bonds and two longer (hydrogen) bonds (Fig. 2.3).

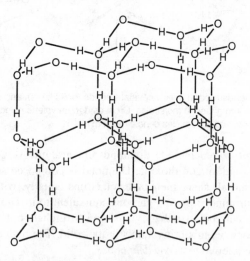

Fig. 2.3 Ice. Long bonds are hydrogen bonds and shorter bonds are covalent bonds.

We have now identified four types of bond:
(1)   *metallic bonds,*
(2)   *ionic bonds* which are due to the attraction of oppositely charged ions,
(3)   *covalent bonds* which are directed in space and involve sharing (as opposed to donation) of valence electrons, and

(4) *hydrogen bonds* which are inter-molecular bonds formed when atoms on different molecules have significantly different partial charges from each other; one of the atoms in a hydrogen bond is H (or D or T).

2.3 Covalent bonds in chemistry

The bonding in molecules is dominated by covalent bonds — in fact there is no such thing as a pure ionic bond since there is some element of electron sharing even in bonds between atoms such as Na and Cl in sodium chloride which is invariably described as ionic. The first realistic description of a covalent bond was given by G. N. Lewis about 1916. Lewis described a bond in terms of *an electron-pair shared between two atoms*; such a bond is often referred to as a *two-centre, two-electron bond*. Some pairs of atoms, such as in O_2 and N_2, have two or three, respectively, two-electron bonds between them. In this case we often represent *each* bond by a line (Fig. 2.4).

Alternatively the valence electrons may be illustrated by dots or crosses, with the bonding electrons located between the atoms they 'glue' together. To aid electron counting, we often draw the electrons from one atom as dots and from the other as crosses, though we know that in the molecule *all electrons are indistinguishable*. In an arbitrary sort of way, the dots and crosses for single bonds are often drawn side-by-side perpendicular to the bond direction but the pairs for double or triple bonds are drawn parallel to the bond direction as shown in Fig. 2.4.

Covalent bonds such as those in CO and the isoelectronic (same number of electrons) N_2 are the strongest bonds known.

a)

N≡N   Li - O - Li

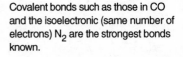

O=O   O=C=O

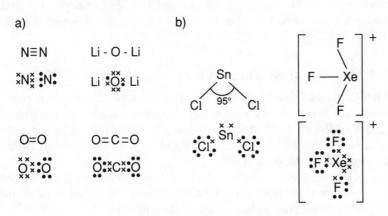

Fig. 2.4 Lewis electron dot and cross structures.

We next address the question of how we could have predicted that N_2 has a triple bond, O_2 a double bond, and F_2 a single bond. Consider the diagrams in Fig. 2.4a and count the number of electrons about each of these atoms. If the bonding electrons are counted as part of each atom's electron count (so are counted twice in the molecule) we find that the count is eight electrons about each atom. This observation gave rise to the *eight electron rule* or *octet rule*. This rule is obeyed strictly on the right-hand side of the second row of the periodic table; it is often obeyed by the same groups in the lower rows of the table. Its justification lies in the following:

(1) a bond requires overlap of orbitals on different atoms, so core orbitals will not make good bonds as they are buried,

(2) second row elements have four valence orbitals, one $2s$ and three $2p$ orbitals,

(3) bonds are stabilizing, so an atom will make as many bonds as possible — this is four two-electron bonds for second row atoms.

The octet rule is inapplicable to the left-hand side of the periodic table since, for example, Li would require *seven* electrons to be added to its valence shell for it to achieve an octet. No bond formation could provide enough energy to stabilize Li^{7-}. If we take an alternative view of the octet rule, we can understand the bonding mode adopted by lithium, namely the formation of ionic compounds using Li^+. When fluorine obeys the eight electron rule it is achieving a full shell of electrons, which we noted in Chapter 1 is a particularly stable configuration. (In fluorine's case its high electronegativity means this is particularly energetically favourable.) Lithium achieves the same end by donating an electron. However, as noted in Chapter 1, it is not correct to say that lithium is more stable as Li^+ (with a full $1s$ orbital) than as Li. It costs energy to remove the electron (the E_i of Li). However, the consequent ionic bond that is formed after lithium loses an electron more than pays for the ionization energy.

The octet rule is also not always observed for elements beyond the second row of the periodic table (Fig. 2.4b) as these elements have more than four valence orbitals available for bonding; further, it is not always clear which electrons are to be counted as valence electrons. For example, we would always expect to find CCl_4, but molecular $SnCl_4$ is unstable with respect to $SnCl_2$ and $Cl_2(g)$.

2.4 Valence shell electron repulsion theory (VSEPR)

After Lewis published his concept of the arrangement of electrons in pairs in a molecule, Gillespie and Nyholm developed the valence shell electron pair repulsion theory (VSEPR) to explain why molecules adopt different geometries. VSEPR theory is very useful as a set of empirical rules to determine molecular geometries.

(1) Once the bonds in a molecule have been formed, place non-bonding (left-over) electrons into lone electron pairs that are localized in space (like bonds but without an atom at the other end).

(2) Find the shape (polyhedron) that minimizes the repulsion between the pairs of electrons, both lone and bonding, so that ligands (the atoms bonded to a central atom) and lone pairs adopt positions that keep them as far from each other as possible; this puts them at the vertices of special geometric shapes (Fig. 2.5).

(3) If there is more than one kind of electron pair the hierarchy of repulsion is:

lone pairs > triple bonds > double bonds > single bonds.

Lone pairs are thought to be most repulsive as they are contracted towards the nucleus and so occupy a greater solid angle than bonding pairs. This means they need to be kept as far apart as possible and

---

Transition metals usually have the atomic electron configuration

$(ns)^2 ([n-1]d)^k$, $k \le 10$.

When transition metals form molecules, the atoms tend to favour an electron count of *eighteen*, resulting from eight electrons in the ns and np orbitals and ten in the $(n-1)d$ orbitals. Transition metals also favour an octahedral geometry with six atoms bonded to the metal in an octahedral shape. It is not always possible for a molecule to have both eighteen electrons and an octahedral geometry, so a variety of shapes result.

Problem: Use the periodic table to determine which of the following molecules follow the eighteen electron rule.

$[ZnCl_4]^{2-}$, $[MnCl_6]^{4-}$, $Fe(CO)_5$, $[Co(NH_3)_6]^{3+}$, $[Cu(H_2O)_6]^{2+}$.

The term *ligand* is used in the broad sense to describe 'surrounding atom or group of atoms'.

occupy the vertices of the polyhedron that have most space about them. Conversely, electronegative ligands require less space as they draw the electron density away from the nucleus so a smaller solid angle is required to accommodate them. In an odd electron system, a single electron may occupy less room than a pair.

The two-dimensional representation of some polyhedra, in particular the pentagonal bipyramid, do not make it obvious which vertex is furthest from its neighbours. Think carefully before assigning lone pairs and atoms to positions.

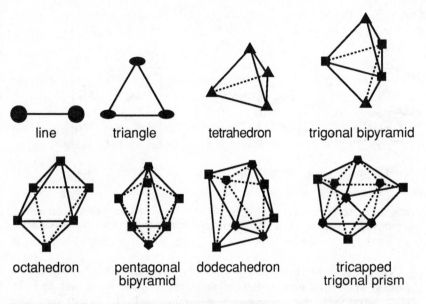

line triangle tetrahedron trigonal bipyramid

octahedron pentagonal bipyramid dodecahedron tricapped trigonal prism

Fig. 2.5 The polyhedra adopted by electron pairs in VSEPR theory. Vertices are represented by shapes indicating the number of edges (lines) meeting there.

VSEPR is very successful in accounting for observed molecular geometries of a wide range of compounds. However, it predicts the wrong geometry for a few systems suggesting that it is in some way deficient. Perhaps its greatest strengths are that it is easy to apply and its failures are easy to remember, so at the very least it is a valuable mnemonic for remembering molecular geometries. Some examples are given in Table 2.1 and others illustrated in Fig. 2.6.

Table 2.1 Experimental geometries and VSEPR predictions (to the nearest degree) for some second row systems.

molecule	experiment	VSEPR	molecule	experiment	VSEPR
H–Be–H	180°	180°	NH_3	107°	109°
H–C–H	102°	120°	BH_3	120°	120°
O=C=O	180°	180°	BF_3	120°	120°
H–O–H	105°	109°	CO_3^{2-}	120°	120°
F–O–F	103°	109°	NH_4^+	109°	109°
NF_3	102°	109°	CH_4	109°	109°

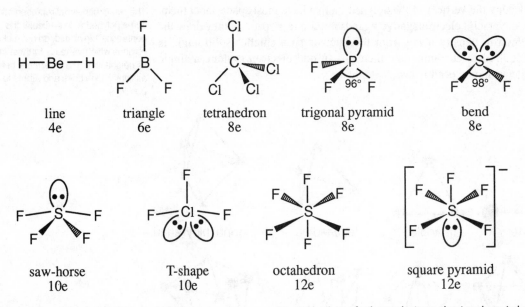

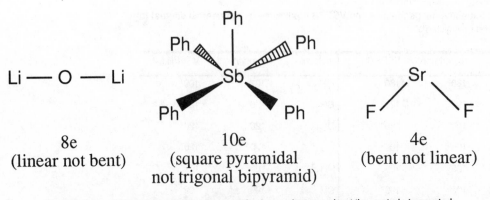

Fig. 2.6 Molecules where VSEPR geometries are the same as experiment. Numbers of valence electrons about each central atom are noted. Lone pairs of electrons are indicated by lobes with dots. PF_3 and SF_2 are distorted tetrahedra, SF_4 and ClF_3 are distorted trigonal bipyramids.

Ni(CO)$_4$ is tetrahedral,
Fe(CO)$_5$ is trigonal bipyramidal,
Co(CO)$_6$ is octahedral.

VSEPR fails for very ionic systems such as the molecule Li_2O, which is formed when solid lithium oxide is vaporized. Li_2O is linear, not bent like H_2O, and is probably more correctly thought of as $Li_2^+O^{2-}$, with repulsion of the positively charged lithium atoms rather than the repulsion of bonding electron pairs dominating. VSEPR, in the form stated above, also has difficulty with many third and fourth row elements, in large part because of the difficulty of deciding how many electrons are to be considered as valence electrons. This problem is particularly acute for transition metal compounds. Problems arise in other systems where the fundamental premise of VSEPR theory, namely that non-bonding electrons are in lone pairs that are localized in space, breaks down. For example, $Sb(C_6H_5)_5$ has a square pyramidal geometry and SrF_2 is bent.

Fig. 2.7 Some VSEPR failures with numbers of valence electrons about the central atom noted.

2.5 Molecular orbital theory

In discussing the Lewis model and VSEPR theory there is a large element of 'it works therefore it is'. Molecular orbital theory (denoted *mo* theory), by way of contrast, is referred to as an *ab initio* (from the beginning) theory. We shall look at a simple version of *mo* theory which assumes that the behaviour of each electron may be described independently of the others by a function called a molecular orbital, or *mo*.

This is exactly how we dealt with many-electron atoms in §1.7; we assumed each electron could be described by an orbital, in that case an atomic orbital, *ao*.

When two or more atoms join together to form a molecule, their electron density is redistributed so that more electron density lies between the atoms to make the bonds. As the atoms within a molecule still resemble the isolated ones, the easiest way to describe the *mos* is in terms of the *aos* of the individual atoms. Further, since the valence *aos* are the ones most involved in any electron redistribution following molecule formation, we shall consider only these orbitals.

Consider one atom approaching another atom along the *z*-axis, each with one *ao* as illustrated in Fig. 2.8. If the *ao* on atom 1 (which is an *s* orbital), ϕ^1, is empty and that on atom 2 (a p_z orbital), ϕ^2, is full, then when ϕ^1 and ϕ^2 overlap, ϕ^1 will share the electron density contained in ϕ^2, and ϕ^2 will share some of the emptiness of ϕ^1.

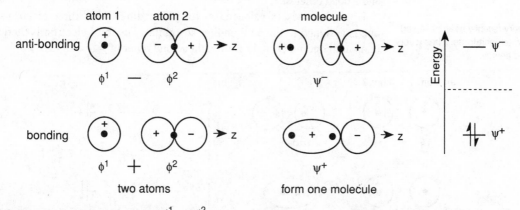

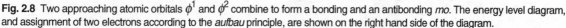

Fig. 2.8 Two approaching atomic orbitals ϕ^1 and ϕ^2 combine to form a bonding and an antibonding *mo*. The energy level diagram, and assignment of two electrons according to the *aufbau* principle, are shown on the right hand side of the diagram.

Mathematically we express what is illustrated in Fig. 2.8 by saying that the two orbitals make a bonding and an antibonding combination, the former concentrating electron density between the atoms and the latter taking it away from the bonding region. In Chapter 1 we used the Greek ϕ for *aos*; now we shall use the Greek ψ (psi) for the *mos*.

In Chapter 1 we drew orbitals with signs indicating the 'phase' of the lobes. This is important for whether a bonding or antibonding orbital is formed. In Fig. 2.8 if the right-hand orbital were reversed the bonding and antibonding combinations would be exchanged.

$$\psi(\text{bonding}) = \psi^+ = \left(\phi^1 + \phi^2\right)$$
$$\psi(\text{antibonding}) = \psi^- = \left(\phi^1 - \phi^2\right)$$

(2.1)

ψ^+ is the more stable orbital as it results in electron density *between* the positive nuclei that reduces the nuclear repulsion and bonds the nuclei together. ψ^+ is therefore called a *bonding orbital*. An electron in ψ^- has its

24 *Molecules: the beginning of chemistry*

electron density outside the bonding area rather than between the atoms. It therefore serves to attract the nuclei outwards as well as not reducing any nucleus–nucleus repulsion. It is therefore called an *antibonding orbital*. We may draw an energy level diagram and allocate electrons according to the *aufbau* principle (§1.7) as is illustrated in Fig. 2.8.

Homonuclear diatomics contain two identical atoms, *e.g.* H_2, O_2, and N_2 with bond orders of respectively one, two, and three.

Problem: The bond energies of H_2^+, H_2, and He_2^+ are 255 kJ mol^{-1}, 432 kJ mol^{-1} and 322 kJ mol^{-1} respectively. Why does the bond order of H_2^+ equal that of He_2^+, but their bond strengths differ significantly? *Hint:* In Chapter 1 we saw that hydrogen-like atoms had more stable energy levels than hydrogen.

He_2 has very recently been observed at very low temperatures. It has a very small bond energy.

First row homonuclear diatomic molecules

The simplest molecule is H_2^+. Each H atom has a $1s$ valence *ao*. As the two atoms are brought together their $1s$ orbitals overlap. Fig. 2.9 illustrates the resulting valence *mo* energy level diagram. For H_2^+, one electron is assigned to the lower *mo*. We define the *bond order* (the number of bonds) to be

$$\tfrac{1}{2}\{(\text{no. of bonding electrons}) - (\text{no. of antibonding electrons})\} \quad (2.2)$$

This definition is in accord with the Lewis concept of a two electron bond. The bond order of H_2^+ is therefore $\tfrac{1}{2}$ (it has half of a two centre, two electron bond).

The next simplest molecule is H_2. H_2 has the same valence *aos* as H_2^+, but has two valence electrons in total. Both electrons are assigned to the bonding ψ^+ orbital with opposite spins. H_2 thus has two bonding electrons and a bond order of 1.

He_2^+ also uses $1s$ valence *aos* for its bonding; it has three electrons, so one is assigned to the antibonding ψ^- orbital. The third electron is therefore in an antibonding orbital resulting in a bond order of $\tfrac{2-1}{2}=\tfrac{1}{2}$.

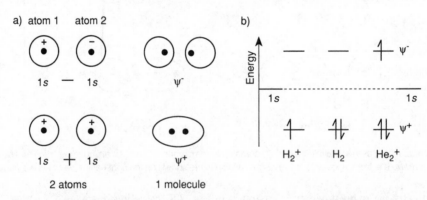

Fig. 2.9 (a) First row diatomic *mo* diagram schematically illustrating the shapes of the orbitals. (b) First row diatomic energy level diagram with electron occupancy of the ground state indicated by arrows. Note that the energy scale is arbitrary and not the same for the three species (the $1s$ orbitals of He are lower than those for H and the number of electrons affects the bond strength).

Sigma and pi orbitals

When one looks down the *bond* (defined to be along the z-axis) of any of the first row diatomic molecules, the electron density appears circular. By analogy with *aos*, *mos* that are circular when viewed down the bond are called sigma, or σ, orbitals. Antibonding sigma orbitals are sometimes denoted σ^*. The *mos* made from *s aos* are always sigma orbitals. When two *p* orbitals approach each other, however, the appearance of the electron density when viewed down the bond axis depends upon the relative orientations of the *aos*.

When two p_z orbitals (*i.e.* ones oriented like the right-hand orbital in Fig. 2.8) approach each other the result is a σ orbital. If the *mo* results from the overlap of two p_x orbitals, then the *mo* looks like a p orbital when viewed down the bond axis and we call it a π orbital (Fig. 2.10).

Second row homonuclear diatomics

The valence orbitals for second row atoms include $2s$ and $2p$ orbitals. Let us assume that an *ao* on atom 1 *only* overlaps with the same orbital of atom 2. The $2s$ orbitals then lead to an energy level diagram just like that of the first row atoms. The $2p$ *aos* give one σ orbital from the overlap of the $2p_z$ *aos*, and two π *mos*, one from the overlap of the $2p_x$ on atom 1 with $2p_x$ on atom 2 and the other from the overlap of $2p_y$ *aos* on each atom. The $2p_z$ *aos* point directly towards each other and so overlap with each other more than the $2p_x/2p_x$ and $2p_y/2p_y$ pairs do, so the amount of bonding (and antibonding) interaction is greater with the $\sigma(2p_z)$ orbital than with the π *mos* as shown in Fig. 2.10. The *mo* energy level diagram for such a molecule is illustrated in Fig. 2.11a. The valence electrons may then be assigned to orbitals according to the *aufbau* principle.

The molecular analogue of the atomic orbital quantum number l is the Greek letter lambda, λ. $\lambda = 0$ for a σ *mo* and $\lambda = 1$ for a π *mo*. The big difference between *ao* and *mo* quantum numbers is that m_λ for $\lambda = 1$ only takes values of $+1$ or -1. The 'missing' $m_\lambda = 0$ orbital is the σ orbital from combining $2p_z$ orbitals, so it has $\lambda = 0$.

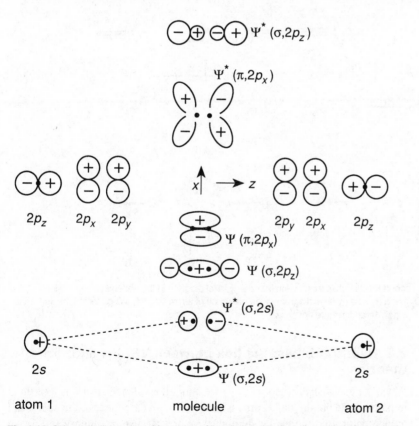

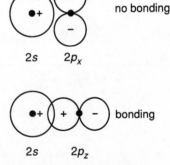

no bonding

bonding

$2p_x$ does not mix with $2s$ because the p orbitals have one half positive and one negative, so the net overlap is zero. $2s$ and $2p_z$ do, however, have net overlap.

Fig. 2.10 Schematic illustration of formation of bonding and antibonding orbitals for second row diatomic molecules. The two *atoms* are shown schematically on either side of the *molecule* that they form. There are $\psi(\pi, 2p_y)$ and $\psi*(\pi, 2p_y)$ orbitals degenerate with the corresponding $\psi(\pi, 2p_x)$ and $\psi*(\pi, 2p_x)$ orbitals but oriented pointing into and out of the page.

The energy order of the orbitals in Fig. 2.11a is only correct for O_2 and F_2. The reason for this is that in determining the *mo* energy level diagram we assumed that a $2p_z$ *ao* on one atom does not overlap with a $2s$ orbital on the other atom. This assumption is not correct (see Fig. 2.8). On the right-hand side of the periodic table we can ignore the error as the higher nuclear charge results in the shielding of the $2p$ electrons by the $2s$ electrons being very effective so there is a large *s/p* energy gap. On the left-hand side of the periodic table, however, the shielding is less effective and significant *s/p* mixing occurs. The result of the *s/p* mixing is that the lower two σ *mos* are lower in energy than Fig. 2.11a indicates and the higher two are even higher. Fig. 2.11b is a better description of $Li_2 - N_2$.

Ne_2 has a very weak bond. It is, however, stronger than that of He_2.

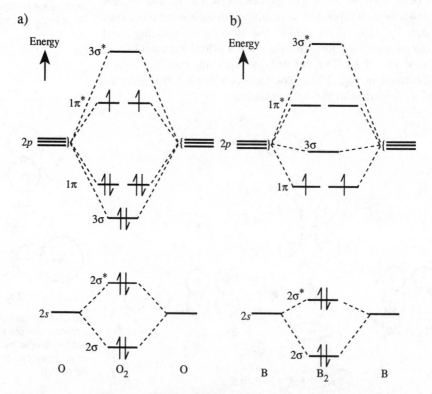

Fig. 2.11 *mo* energy level diagram for second row diatomic molecules: (a) energy level diagram in the absence of *s/p* mixing (showing the electron assignment for O_2); and (b) energy level diagram in the presence of *s/p* mixing (showing electron assignment for B_2). σ and π orbitals are numbered in order, 1σ and $1\sigma^*$ are not shown.

2.6 Hybrid orbitals: the link between VSEPR and *mo* theories

When we use VSEPR theory (§2.4) we count the valence electrons, put them in pairs, some bonding and some non-bonding, that are localized in space and thence determine molecular geometry. When we use *mo* theory (§2.5) we describe the behaviour of each electron by an orbital that is spread out over the whole molecule. How *we* choose to *describe* the electrons does not affect the fact that molecules form because electron density is concentrated between atoms. However, how we describe the electrons does affect our understanding of the molecules and our ability to predict their chemistry.

In most cases, the concept of each bond being localized between two atoms with two electrons in it is a good description of what happens. Certainly it works for diatomic molecules where there are only two atoms. In some cases, however, it is too simplistic. Benzene, for example, has six identical carbon atoms and six identical C–C bonds whose length is midway between our expectation for a single bond and a double bond. One description of the bonding in benzene is to describe it as a *resonance hybrid* of the two bonding arrangements shown in Fig. 2.12. Benzene is also much less reactive than might be expected by analogy with other molecules with double bonds. A different representation where we replace the alternating double/single bonds by a circle, which reminds us that the carbon–carbon bonds in benzene have bond orders of 1.5 and the π electron density is in some way delocalized over the molecule, is sometimes helpful.

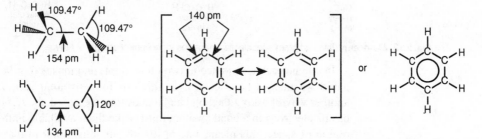

Fig. 2.12 Ethane, ethene, and benzene.

A second problem that occurs when we look at benzene is that there is no obvious correlation between the valence $2s$ and $2p$ orbitals on each carbon and the angles between the bonds that hold the molecule together. In particular, the three $2p$ orbitals are oriented at 90° to one another. The s orbital is fully occupied in atomic carbon thus not able to share electrons from another atom and is in any case partly buried beneath the p orbitals. So how do we make bond angles of 120°? Similarly, how can we account for the four 109.47° bond angles observed for the tetrahedral methane molecule?

Before answering these questions, let us first consider the bonding in C_2 in a little more detail. The left-hand side of Fig. 2.13 shows the allocation of electrons to orbitals for this molecule. The $2s$ *aos*, with some contribution from the $2p_z$ orbitals, of each C form a bonding and an antibonding orbital that are both fully occupied so have little net bonding effect. The $2p$ *aos* lead to an occupied σ bonding orbitals and two π bonding orbitals that are both occupied. So we have three bonding orbitals and one antibonding orbital fully occupied. The net result is a bond order of $(3 - 1) = 2$.

The antibonding electrons in C_2 reduce the maximum possible bond count of each carbon by one. The electron density in the antibonding orbital is outside the C–C bond area. When ethyne is formed, the hydrogen atoms that are added to C_2 use this antibonding electron density to make C–H bonds since it is in precisely the correct place for such bonds. This is illustrated in Fig. 2.13.

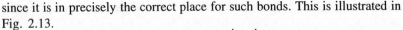

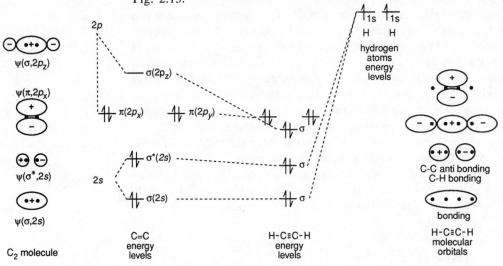

Fig. 2.13 *Mo* energy level diagram and electron densities for occupied orbitals of ethyne.

In ethyne each C is joined to two other atoms and makes four bonds. How could we instead bond each C to three or four different atoms? (This is another way of asking the bond angle questions we posed above.) In order to use its electrons in a bond an atom must make them available in the bonding region of space, this means one of its orbitals must point in that direction. We may describe how carbon does this using the concept of *hybrid aos*. Hybrid *aos* are atomic orbitals that are a mixture of, *e.g.*, 2*s* and 2*p* *aos*. The valence electrons of atomic carbon are as illustrated in Fig. 2.14a; there are 2*s* electrons and two 2*p* electrons thus leaving one 2*p* orbital unoccupied. To make hybrid *aos* we proceed as follows.

(1) Put the correct amount of energy into the atom to excite one of the 2*s* electrons into the unoccupied 2*p* orbital. There is now one electron in each of the 2*s* and 2*p* orbitals, and the total electron density is spherical.

(2) Identify new sets of four orbitals, the hybrid orbitals (whose combined volume is the same as the four 2*s* and 2*p* orbitals), and whose orientation allows for non-linear molecules. Each hybrid is allocated one electron and so they are prepared to share electrons in covalent bonds.

Three sets of hybrid orbitals are commonly used for carbon (and other second row atoms):

*   four *sp*^{3} hybrid orbitals (*i.e.* each orbital has $\frac{3}{4}$ *p* character and $\frac{1}{4}$ *s* character) that are oriented tetrahedrally with respect to one another, to make four σ bonds;
*   three *sp*^{2} hybrid orbitals (*i.e.* each orbital has $\frac{2}{3}$ *p* character and $\frac{1}{3}$ *s* character) plus one *p* orbital, to make three σ bonds and one π bond;
*   two *sp* hybrid orbitals (*i.e.* each orbital has $\frac{1}{2}$ *p* character and $\frac{1}{2}$ *s* character) and two *p* orbitals, to make two σ bonds and two π bonds.

Atoms which have valence *d* orbitals can have hybrid orbitals with some *d* character. For example, as noted in Fig. 2.6, SF_6 has six pairs of valence electrons all in equivalent bonds. The hybridization scheme *sp*^{3}*d*^{2} on S accounts for this.

Thus ethane carbons are *sp*^{3} hybridized, ethene carbons are *sp*^{2} hybridized, and ethyne carbons are *sp* hybridized.

Towards the bottom of the periodic table, the bond energy is not sufficient to pay the hybridization price and molecules such as $SnCl_2$ are found.

The orbital arrangements with each hybridization set are illustrated in Fig. 2.14. The energy price for the hybridization is compensated by the additional three, two, or one bonds it can now make.

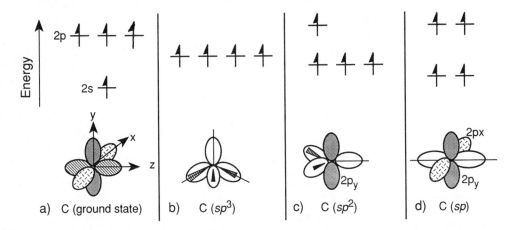

Fig. 2.14 (a) Ground valence electron configuration for atomic carbon (*p* electron densities are shaded to show which lobes belong to a single orbital); electron configuration and bonding arrangements for (b) sp^3 hybridized C, (c) sp^2 hybridized C, and (d) sp hybridized C. There should be a little lobe on the other side of the carbon atom for each hybrid due to the contribution from the 'back' lobes of the *p* orbitals. This is not illustrated for clarity. (Note. These sketches are representations of the electron densities, ψ^2, in an orbital rather than the orbital, ψ, in contrast to previous diagrams.)

2.7 Molecular fingerprints: using the electromagnetic spectrum

In the preceding sections of this chapter we have examined how atoms are held together in molecules. But how can we tell that §2.1 – §2.6 represent the true shapes of molecules? As discussed in Chapter 1 molecules are much too small for us to see ourselves, so we usually use a variety of wavelengths of electromagnetic radiation (Fig. 2.15) and different instruments (spectrometers) to provide us with the means to 'see' molecules.

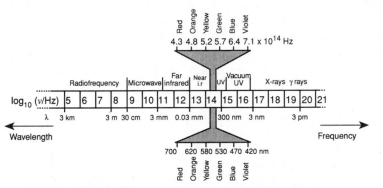

Fig. 2.15 The electromagnetic spectrum with the visible region expanded.

We have already discussed (§1.3 – §1.7) the emission spectra we can measure for atomic hydrogen, hydrogen-like atoms, and many-electron atoms such as lithium. The lines of energy emitted corresponded exactly with the energy gap between different atomic electron energy levels. Thus the spectra are fingerprints for the identity of the atoms. When we come to study molecules, we more commonly measure the amount of energy the molecule needs to *absorb* to jump *from* the ground state *to* a higher state (rather than the emission from the higher state). Table 2.2 lists a number of different spectroscopies, the radiation they use, and the molecular features they probe. Each spectrum contributes to a fingerprint for a molecule. With an unidentified molecule, our task is to interpret the fingerprints to give us information we require about its bonding, geometry, and chemistry. We shall look at the most commonly used spectroscopic techniques in some more detail.

The use of spectroscopy is by no means limited to the chemistry laboratory. Most kitchens are now equipped with microwave ovens, which make use of the fact that microwave radiation causes water molecules to rotate thus warming up the food. Other molecules with similar rotational energies are also heated up in microwave ovens.

Table 2.2 Different spectroscopic techniques and the radiation they employ.

Type of spectroscopy	Radiation	Causes changes in
nuclear magnetic resonance	radiowave	proton spin
rotational	microwave	molecular rotations
electron spin resonance	microwave	electron spin
vibrational	infrared	molecular vibrations
absorbance	visible / ultraviolet	valence electrons
photoelectron spectroscopy	far UV / X-ray	core electrons
Mossbauer	γ-rays	nuclear structure

Photoelectron spectroscopy

We have already considered photoelectron spectroscopy (PES) of atoms in the context of the development of quantum mechanics. In measuring a PES spectrum of a molecule, we measure the kinetic energy of electrons ejected from a sample of molecules when high energy photons hit the sample. The energy difference between the incident radiation and the kinetic energy of the emitted electron tells us how much energy was required to remove an electron from a *mo*. Thus PES is an experimental means of determining a *mo* energy level diagram. The PES for N_2 is shown in Fig. 2.16. This spectrum is historically important for PES since prior to the PES spectrum being measured it had been thought that N_2 and O_2 had the same energy order of *mos* (Fig. 2.11a).

NMR spectroscopy has become an invaluable clinical tool for imaging internal organs, when cleverly implemented and relabelled magnetic resonance imaging.

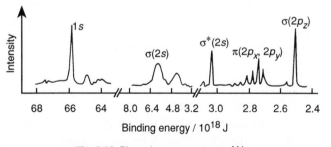

Fig. 2.16 Photoelectron spectrum of N_2.

UV / visible absorbance spectroscopy

Ultra-violet (UV) and visible radiation have the correct energies to move electrons from ground state energy levels to higher ones. As well as telling us about how the electron density of a molecule is changed when electrons are excited, we can also use UV/visible absorption spectroscopy to tell us how systems are changing. An example is given in Fig. 2.17.

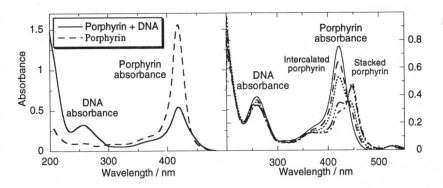
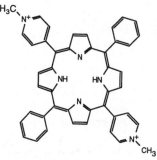

Fig. 2.17 UV / visible absorption spectra of the porphyrin illustrated. Left-hand graph shows 6 μM porphyrin in water with and without calf thymus DNA (40 μM). The small shift to longer wavelengths and reduced absorption intensity when DNA is added is characteristic of a molecule binding sandwiched (or intercalated) between DNA base pairs. The right-hand graph shows the effect of increasing NaCl (0, 20, 60, 80, 120, 140, 180 mM) on a solution containing 5 μM porphyrin and 40 μM DNA . In this case we are seeing the shift from an intercalative binding mode at 0 mM NaCl to one where the porphyrin is externally stacked on the DNA at high NaCl concentration.

NMR spectroscopy

Nuclear magnetic resonance (NMR) spectroscopy is the most widely used structural technique in chemistry laboratories. It is routinely used in an empirical manner following a few simple rules. However, it is perhaps the most difficult of the spectroscopic techniques to explain. The basis of NMR lies in the fact that many atomic nuclei possess spin in much the same way as electrons do.

The hydrogen nucleus, ^{1}H, has a spin of magnitude $\frac{1}{2}$ and, since it is charged, acts as a magnet. When an external magnetic field is imposed on a hydrogen atom, its spin orients either parallel (denoted α) to the magnetic field or antiparallel to it (denoted β). The two orientations have different energies with α being lower and β being higher than the levels with no magnetic field. This creates two energy levels (Fig. 2.18).

In ^{1}H NMR spectroscopy we measure the energy required to transfer a hydrogen nucleus from the α state to the β state. The energy gap is dependent upon how much of the magnetic field the nucleus 'feels'. Since the nucleus is shielded from an imposed magnetic field by the surrounding electrons in the atom or molecule, we can use the energy of the transition to determine the environment (neighbouring atoms in the molecule) of a hydrogen. Typical relative positions for different functional groups are illustrated in Fig. 2.18.

To illustrate NMR spectroscopy stand up and spin round on the spot. Now try to walk in a smallish clockwise circle at the same time. Next try to walk in an anticlockwise circle. Depending on your direction of spin, one of the combinations will be much harder to do. You have found two different energy states.

Compass needles in the earth's magnetic field act just like protons in a magnetic field: the needle usually points towards north, which is its most stable alignment, but it can get stuck pointing south in a metastable position.

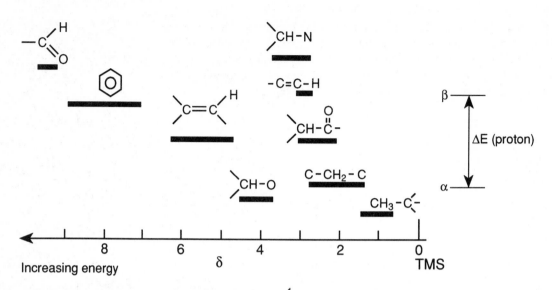

Fig. 2.18 Typical relative positions for different functional groups in a ^{1}H NMR spectrum. The energy differences between the α and β states are very small. The differences between the transition energies for different type of protons are even smaller. We usually use an energy scale relative to the standard tetramethylsilane, Si(CH$_3$)$_4$, (TMS): $\delta = \dfrac{\Delta E(\text{proton}) - \Delta E(\text{TMS proton})}{\Delta E(\text{TMS proton})} \times 10^6$.

All nuclei with an odd number of protons plus neutrons (the mass number A) have a non-zero nuclear spin. Some atoms with A even, such as D, also have a non-zero nuclear spin; for D it is 1. The NMR spectroscopy of nuclei with nuclear spin equal to $\frac{1}{2}$, such as ^{13}C, ^{19}F and ^{31}P, are in principle very similar to that of hydrogen.

Let us consider an example. $CH_3CH_2CH_2NO_2$ has three different types of hydrogens which absorb radiowaves of different frequencies to change their spins. This results in the NMR spectrum illustrated in Fig. 2.19. The CH$_3$ hydrogens have most electron density about them, so occur at lowest energy; the neighbouring CH$_2$ hydrogens occur at a slightly higher energy because a bonded carbon removes more of the protecting electron density than does an H. The highest energy band corresponds to the CH$_2$ next to the electronegative NO$_2$ group. *The areas under the bands are proportional to the number of hydrogens whose transitions occur at that energy.*

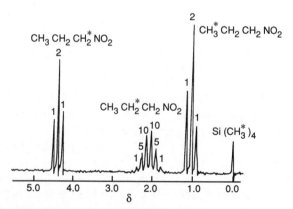

Fig. 2.19 ^{1}H NMR spectrum of CH$_3$CH$_2$CH$_2$NO$_2$. Tetramethylsilane (TMS), which has 12 equivalent hydrogens is included as a standard.

Another extremely useful feature in NMR spectra is hydrogen–hydrogen coupling. Chemically equivalent protons such as the three protons of the freely rotating methyl group in ethanol show no coupling to each other (it does occur but has no net effect in the spectrum). All other protons that are close to each other couple and split the lines in an NMR spectrum. If there are two identical protons near one we are interested in, then the spectrum of our proton will be split into (2+1) peaks whose relative magnitudes are 1–2–1. *The relative intensities of the component peaks of a multiplet indicate the number of (more-or-less) equivalent neighbouring hydrogens of any given hydrogen.* The spectrum of Fig. 2.19 illustrates the effect of coupling. The terminal CH_3 hydrogen atoms and those in the CH_2 adjacent to the NO_2 group are approximately equivalent so give rise to the observed 1 5 10 10 5 1 splitting pattern.

number of neighbouring protons	relative intensities of peaks
0	1
1	1 1
2	1 2 1
3	1 3 3 1
4	1 4 6 4 1
5	1 5 10 10 5 1
	etc.

Pascal's triangle summarizes the relative intensities of the members of a 'couplet'.

Vibrational spectroscopy

The final type of spectroscopy we shall discuss is vibrational or infrared spectroscopy. The energy required to cause a molecule to vibrate by stretching or bending its bonds is provided by infrared radiation. Precisely what frequency is absorbed gives us information about how strong a bond is, which in turn often enables us to identify the different types of bonds a molecule contains. Infrared spectroscopy is thus often used to provide a fingerprint of a molecule or to identify a newly synthesized compound.

In a symmetric molecule such as O=C=O, the two C=O bonds do not vibrate independently, instead we find four vibrations: two (degenerate) bends, a symmetric stretch, and an antisymmetric stretch.

O = C = O   bend

← O = C = O →   O ═ C ═ O
symmetric stretch

O = C = O →   O═ C ═ O
asymmetric stretch

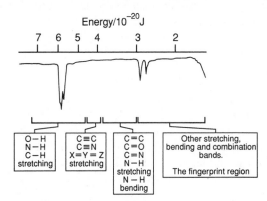

Fig. 2.20 Spectrum of the liquid paraffin Nujol (contains only C–H and C–C bonds) and characteristic energies of some vibrations.

Infrared spectroscopy is easy to understand if we imagine bonds are springs between atoms. Then the stronger the bond the greater the amount of energy we need to use to cause that bond to vibrate. The same analogy leads us to conclude (correctly) that we would expect bond bending vibrations to require much lower energy than bond stretching vibrations. Fig. 2.20 indicates some typical vibrational transition energies.

The many spectroscopic techniques listed in Table 2.2 enable us to test experimentally the theory that we used earlier in this chapter to describe molecules: the beginning of chemistry. In the next chapters we look at why and how molecules react.

3 Chemical energetics

3.1 Introduction

In the last two chapters we considered the structure and properties of atoms and molecules. Next we begin to provide a basis for understanding chemical reactivity — for predicting whether a molecule can react (by itself or with another molecule), and if it does, what products can be formed. For example we might wish to appreciate why gaseous nitrogen trifluoride is stable with respect to its elements at 298 K:

$$NF_3(g) \overset{\times}{\longrightarrow} \tfrac{1}{2}N_2(g) + \tfrac{3}{2}F_2(g) \tag{3.1}$$

whereas liquid nitrogen trichloride is spontaneously explosive under the same conditions:

$$NCl_3(\ell) \longrightarrow \tfrac{1}{2}N_2(g) + \tfrac{3}{2}Cl_2(g) \tag{3.2}$$

The latter example reminds us that many chemical reactions are accompanied by the evolution of significant amounts of heat. We begin by considering this aspect.

3.2 Enthalpy and chemical reaction

When one mole of NCl_3 decomposes to form its constituent elements, N_2 and Cl_2 according to eqn (3.2), 231 kJ of energy is released in the form of heat. It is useful to introduce the idea of a *system* and its *surroundings*. In the present case the former consists of one mole of liquid NCl_3 in its container; the latter comprises the laboratory in which the explosion occurs together with the rest of the universe! The explosion is illustrated in Fig. 3.1.

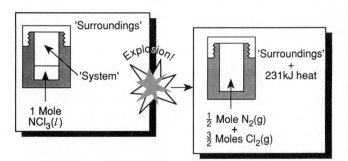

Fig. 3.1 Explosive reaction of nitrogen trichloride.

It is further helpful to develop the concept of the *enthalpy* of a chemical system. This is denoted by the symbol H. The amount of heat absorbed or released in a chemical reaction, *if measured at constant pressure*, is equal to the *change in enthalpy* of the system. In the explosion of Fig. 3.1 the system has lost enthalpy; the reaction is *exothermic*. We denote this as

$$\Delta H = -231 \text{ kJ mol}^{-1} \qquad (3.3)$$

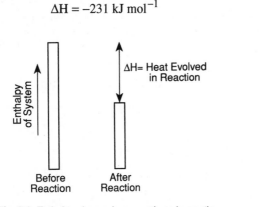

Fig. 3.2 Enthalpy change in an exothermic reaction.

The negative sign in eqn (3.3) indicates that the system has lost enthalpy and the symbol Δ (Greek 'delta') signifies 'change in'. For an *endothermic* process ΔH is positive. For example, breaking a C–H bond in methane to form a methyl radical

$$CH_4(g) \rightarrow CH_3^{\bullet}(g) + H^{\bullet}(g) \qquad \Delta H = +416 \text{ kJ mol}^{-1} \qquad (3.4)$$

absorbs over 400 kJ of heat per mole of methane. We shall see later in this chapter that enthalpy changes are a very useful, albeit partial, guide to chemical reactivity.

The observant reader will notice that Fig. 3.1 shows the explosion of NCl_3 under conditions of constant volume rather than constant pressure — the container has been given a tightly fitting lid to limit the consequences of the explosion. Consequently the heat evolved is not exactly 231 kJ. However, the difference — approximately 5 kJ — is sufficiently small that we can safely neglect the difference between the enthalpy change and the change in 'internal energy'. The former is the heat evolved at constant pressure (ΔH) whilst the latter is the heat evolved at constant volume (denoted ΔU).

We use the notation $CH_3^{\bullet}$ when we wish to emphasize that there is an unpaired electron present. Such a moiety is called a radical.

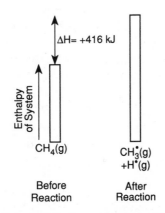

Fig. 3.3 An example of an endothermic reaction.

3.3 Hess' law

Experiments on a wide range of chemical reactions led the chemist G. H. Hess to observe in 1840 that: *if a reaction is conducted in more than one stage, the overall enthalpy change is the sum of the enthalpy changes involved in the separate stages*. This is now known as Hess' law and can be

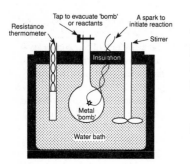

The figure above shows how to measure the enthalpy of reaction. The slight temperature rise produced is measured accurately and the energy released determined by measuring the amount of electrical energy which has to be put into the system – when no reactants are present – to produce an equivalent temperature rise.

A simple statement of the first law of thermodynamics is that energy can neither be created nor destroyed.

illustrated by the following example. Consider the formation of gaseous hydrogen chloride from its constituent elements at 298 K:

$$H_2(g) + Cl_2(g) \rightarrow 2HCl(g) \qquad \Delta H = -184 \text{ kJ mol}^{-1} \qquad (3.5)$$

This is a strongly exothermic reaction. It might (at least in principle) be imagined to be brought about in the following separate stages, each at 298 K.

(1) Hydrogen is dissociated into atoms:

$$H_2(g) \rightarrow 2H(g) \qquad \Delta H = +436 \text{ kJ mol}^{-1} \qquad (3.6)$$

(2) Chlorine molecules are likewise fragmented:

$$Cl_2(g) \rightarrow 2Cl(g) \qquad \Delta H = +242 \text{ kJ mol}^{-1} \qquad (3.7)$$

(3) The different atoms are allowed to recombine:

$$H(g) + Cl(g) \rightarrow HCl(g) \qquad \Delta H = -431 \text{ kJ mol}^{-1} \qquad (3.8)$$

It can be seen that

$$-184 = 436 + 242 - (2 \times 431) \qquad (3.9)$$

so confirming the validity of Hess' law.

A little thought shows that Hess' law is simply a statement of the law of the conservation of energy phrased in a chemically pragmatic form. Imagine a hypothetical reaction,

$$A \rightarrow C \qquad \Delta H = \alpha \qquad (3.10)$$

but which might also happen *via* an intermediate species, B:

$$A \rightarrow B \quad \Delta H = \beta; \quad B \rightarrow C \quad \Delta H = \gamma \qquad (3.11)$$

and suppose that $\alpha > \beta + \gamma$. Next imagine that the following sequence of reactions could be brought about:

$$A \rightarrow B \rightarrow C \rightarrow A \qquad (3.12)$$

Now the chemistry of eqn (3.12) has resulted in no net change — we started and finished with one mole of A. But, by eqns (3.10) and (3.11), $\Delta H = \alpha + \beta + \gamma$ for this series of reactions, which is negative. Thus energy, in the form of heat, would be created from nothing. Alternatively if $-\alpha < \beta + \gamma$ then the cycle

$$A \rightarrow C \rightarrow B \rightarrow A \qquad (3.13)$$

again results in the creation of energy without cost! Experience tells us that neither of these possibilities can be realistic, so Hess' law stands and

$$\alpha = \beta + \gamma \quad \text{and} \quad \Delta H = 0 \qquad (3.14)$$

3.4 The standard enthalpy of formation

The standard enthalpy of formation of a compound, ΔH_f^O *is the enthalpy change involved in the formation of one mole of the substance from its elements, each element being in its usual form at one atmosphere pressure and the specified temperature.* In this book ΔH_f^O means ΔH_f^O at 298 K. Temperature *and* pressure are specified since both these factors influence the enthalpy of a substance. Representative values of ΔH_f^O are listed in Table 3.1.

The states such as $O_2(g)$, $Hg(\ell)$, and Fe(s) in which we find elements at SATP therefore have zero standard enthalpy of formation.

Table 3.1 Standard enthalpies of formation.

Substance	ΔH_f^o / kJ mol^{-1}	Substance	ΔH_f^o / kJ mol^{-1}	Substance	ΔH_f^o / kJ mol^{-1}
$H_2O(g)$	−242	$Ca(OH)_2$	−987	$PF_5(g)$	−1594
$H_2O(\ell)$	−286	$CH_4(g)$	−75	$SO_2(g)$	−297
$CO_2(g)$	−393	$C_2H_6(g)$	−85	$SO_3(g)$	−395
$CO(g)$	−111	$C_4H_{10}(g)$	−126	$H_2S(g)$	−23
$NF_3(g)$	−125	$C_2H_4(g)$	+52	$Br_2(g)$	+32
$NCl_3(\ell)$	+230	$C_2H_2(g)$	+227	$Br_2(\ell)$	0
$CaO(s)$	−636	$C_6H_6(\ell)$	+52	$Br_2(s)$	−11

Notice that in Table 3.1 where ΔH_f^o values are cited for the same compound, notably bromine, in different physical states (solid, liquid or gas) it is possible to evaluate

the enthalpy of vaporization $\qquad \Delta H_{vap}^o = \Delta H_f^o(g) - \Delta H_f^o(\ell) \qquad$ (3.15)

the enthalpy of fusion $\qquad \Delta H_{fus}^o = \Delta H_f^o(\ell) - \Delta H_f^o(s) \qquad$ (3.16)

and the enthalpy of sublimation $\qquad \Delta H_{sub}^o = \Delta H_f^o(g) - \Delta H_f^o(s) \qquad$ (3.17)

The data shown in Table 3.1 enables us to calculate enthalpy changes for a wide range of important reactions by simply invoking Hess' law. Some examples are as follows.

- The standard enthalpy of combustion of butane, written ΔH_{comb}^o, may be represented diagramatically as

$$C_4H_{10}(g) + \tfrac{13}{2}O_2(g) \xrightarrow{\ \Delta H_{comb}^o\ } 4CO_2(g) + 5H_2O(\ell) \qquad (3.18)$$

$$\Delta H^o = -\Delta H_f^o(C_4H_{10}) \qquad \qquad \Delta H^o = 4\Delta H_f^o(CO_2) + 5\Delta H_f^o(H_2O)$$

$$4C(s) + 5H_2(g) + \tfrac{13}{2}O_2(g)$$

The diagram used to calculate the enthalpy of combustion of butane can be replaced by the following algebra:

$$\Delta H_{comb}^o$$
$$= 4(-393) + 5(-286) + 126$$
$$= -2877 \text{ kJ mol}^{-1}$$

Hence,

$$\Delta H_{comb}^o = 4\Delta H_f^o(CO_2(g)) + 5\Delta H_f^o(H_2O(\ell)) - \Delta H_f^o(C_4H_{10}(g)) \qquad (3.19)$$

- The enthalpy of hydrogenation of ethene, ΔH_{hydn}^o, may similarly be represented as

$$C_2H_4(g) + H_2(g) \xrightarrow{\ \Delta H_{hydn}^o\ } C_2H_6(g) \qquad (3.20)$$

$$\Delta H^o = -\Delta H_f^o(C_2H_4) \qquad \qquad \Delta H^o = \Delta H_f^o(C_2H_6)$$

$$2C(s) + 3H_2(g)$$

Algebraically, the enthalpy of the hydrogenation of ethene

$$\Delta H_{hydn}^o = -85 - 52$$
$$= -137 \text{ kJ mol}^{-1}$$

Thus, Hess' law requires that,

$$\Delta H^o_{hydn} = \Delta H^o_f\left(C_2H_6\right) - \Delta H^o_f\left(C_2H_4\right) \qquad (3.21)$$

Notice two features in the above calculations.

Although $\Delta H^o_f(O_2)$ is zero,

$$\Delta H^o_f(O_3) = 143 \text{ kJ mol}^{-1}$$

since it is not the usual form of oxygen at SATP.

(1) Both $\Delta H^o_f(O_2(g))$ and $\Delta H^o_f(H_2(g))$ are zero. This follows directly from the definition of ΔH^o_f given above since $O_2(g)$ and $H_2(g)$ are the usual forms of these elements at one atmosphere pressure and 298 K.

(2) Since the values of ΔH^o_f relate to standard conditions (SATP), so too do the values of ΔH^o_{comb} and ΔH^o_{hydn} evaluated using the values from Table 3.1.

3.5 ΔH^o_f and the chemistry of carbon

Chapter 1 described some of the forms in which elemental carbon can be found. The most common allotropes are graphite and diamond for which

$$\Delta H^o_f(\text{graphite}) = 0 \qquad (3.22)$$

$$\Delta H^o_f(\text{diamond}) = +1.9 \text{ kJ mol}^{-1} \qquad (3.23)$$

showing that graphite is the more thermodynamically stable form of carbon at 298 K and one atmosphere pressure since the transformation

The qualitative conclusion about the relative stability of diamond and graphite remains unchanged when entropy effects — as introduced later in this chapter — are included in the discussion.

$$C(\text{diamond}) \rightarrow C(\text{graphite}) \qquad \Delta H^o = -1.9 \text{ kJ mol}^{-1} \qquad (3.24)$$

is exothermic.

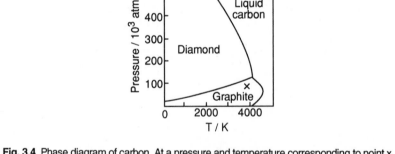

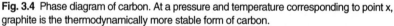

Fig. 3.4 Phase diagram of carbon. At a pressure and temperature corresponding to point x, graphite is the thermodynamically more stable form of carbon.

Chemists use *phase diagrams* to summarize the conditions (of temperature and pressure) under which different phases of a substance are the most stable form. The axes of the diagram are pressure and temperature and the zones in which particular phases are the most stable are labelled. Fig. 3.4 shows the phase diagram of carbon from which it can be seen that at SATP graphite is indeed the stable form. Note however that at suitably high pressures and relatively low temperatures diamond is the more stable phase, whilst at sufficiently large temperatures both solid forms will melt to form liquid carbon.

3.6 Bond enthalpies

Consider the dissociation reaction

$$O_2(g) \rightarrow O(g) + O(g) \qquad \Delta H^o = +490 \text{ kJ mol}^{-1} \qquad (3.25)$$

in which one mole of oxygen molecules form two moles of oxygen atoms. The value of ΔH^o for the process can be taken as a measure of the strength or energy of the O=O bond. The corresponding value for the dissociation of chlorine

$$Cl_2(g) \rightarrow Cl^{\bullet}(g) + Cl^{\bullet}(g) \qquad \Delta H^o = +238 \text{ kJ mol}^{-1} \qquad (3.26)$$

shows that the Cl–Cl bond is relatively weaker than the O=O bond.

It is interesting to develop the above and ask how strong are the chemical bonds between the different atoms in molecules. For example the bond enthalpy of a C–H bond, D(C–H), can be assumed to be related to the standard enthalpy change, ΔH^o, for the reaction

$$CH_4(g) \rightarrow C(g) + 4H(g) \qquad (3.27)$$

If the four bonds are assumed to be identical then

$$D(C-H) = \tfrac{1}{4}\Delta H^o \qquad (3.28)$$

The value of ΔH^o for eqn (3.27) is obviously difficult to measure directly but can be estimated using a Hess' law calculation based on the following three reactions for which the enthalpy change is experimentally accessible:

$$H_2(g) \rightarrow 2H(g) \qquad \Delta H^o = +436 \text{ kJ mol}^{-1}$$

$$C(s) \rightarrow C(g) \qquad \Delta H^o = +717 \text{ kJ mol}^{-1} \qquad (3.29)$$

$$C(g) + 2H_2(g) \rightarrow CH_4(g) \qquad \Delta H^o = -75 \text{ kJ mol}^{-1}$$

The following *Hess cycle* can be drawn

$$
\begin{array}{ccc}
CH_4(g) & \xrightarrow{\Delta H^o = ?} & C(g) + 4H(g) \\
\Big\downarrow \Delta H^o = 75 \text{ kJ} & & \Big\uparrow \Delta H^o = 2 \times 436 \text{ kJ} \\
C(s) + 2H_2(g) & \xrightarrow{\Delta H^o = 717 \text{ kJ}} & C(g) + 2H_2(g)
\end{array}
$$

The enthalpy change for the reaction of eqn (3.27) is

$$\Delta H^o = +1664 \text{ kJ mol}^{-1} \qquad (3.30)$$

from which it may be deduced that

$$D(C-H) = +\tfrac{1}{4}(75 + 717 + 2 \times 436) \text{ kJ mol}^{-1} \qquad (3.31)$$

$$= +416 \text{ kJ mol}^{-1}$$

We may then evaluate the bond enthalpy of a C–C bond, D(C–C), by considering the reaction

$$C_2H_6(g) \rightarrow 2C(g) + 6H_2(g) \qquad (3.32)$$

for which

$$\Delta H^o = 6D(C–H) + D(C–C) \qquad (3.33)$$

Cl_2 contains a single bond whereas O_2 has a double bond.

D(C–H) represents the *average* bond enthalpy. Experiments show that this is in fact close to, but not exactly, the enthalpy change involved in the single dissociation

$CH_4(g) \rightarrow CH_3(g) + H(g)$

for which ΔH^o is +425 kJ mol^{-1}. The corresponding values for the successive losses of H atoms from CH_3, CH_2 and CH are +445, +445 and +340 kJ mol^{-1} respectively.

It follows that we can use our knowledge of D(C–H) to find D(C–C) if ΔH° can be found. This again can be estimated using a Hess cycle:

$$C_2H_6(g) \xrightarrow{\Delta H^\circ = ?} 2C(g) + 6H(g)$$

$$\Delta H^\circ = -\Delta H_f^\circ(C_2H_6) = 85 \text{ kJ}$$

$$\Delta H^\circ = 3 \times 436 \text{ kJ}$$

$$2C(s) + 3H_2(g) \xrightarrow{\Delta H^\circ = 2 \times 717 \text{ kJ}} 2C(g) + 3H_2(g)$$

Problem: Determine D(C–C) in the manner outlined in the text.

The result is that D(C–C) has a value of $+331 \text{ kJ mol}^{-1}$.

If this exercise is continued for a wide range of molecules it is possible to systematically build up a table of the bond enthalpies for a wide range of chemical bonds. Some typical data are shown in Table 3.2.

The values for diatomic molecules in Table 3.2 are accurate. However, those for polyatomic species are less so, since, as illustrated in a previous marginal note for the case of a C–H bond in CH_4, CH_3 *etc.*, an individual bond enthalpy may vary somewhat according to the precise structure of the individual molecule in which the bond is located. Nevertheless the values give good rough estimates for bond strengths that are likely to occur.

Table 3.2 Some Bond Enthalpies.

Bond	$D / \text{kJ mol}^{-1}$	Bond	$D / \text{kJ mol}^{-1}$
F–F	157	C–Cl	339
Cl–Cl	243	C–F	485
Br–Br	194	C–O (general)	358
I–I	153	C=O (ketones)	745
H–H	436	C–H (general)	416
O=O	498	C–C (general)	346
N≡N	945	C=C (general)	609
H–F	569	C≡C (general)	835
H–Cl	432	O–O (H_2O_2)	146
H–Br	366	N–N (N_2H_4)	163
H–I	299	N–F	278
O–H (H_2O)	464	N–Cl	192

Table 3.2 and related data show some trends that are easy to understand and others that are unexpected.

*   Single bonds are generally weaker than double bonds which are weaker than triple bonds.

Note that H and He comprise the first row of the periodic table. The second row is Li – Ne.

*   In any (vertical) group of the periodic table, A_2, A_3, A_4, ..., where the subscript denotes the row of the periodic table, $D(A_i–X)$ decreases as A_i gets heavier when there are no lone pairs on X. Thus

$$D(F–H) > D(Cl–H) > D(Br–H) > D(I–H)$$

Similar trends are seen for (O–H, S–H, Se–H and Te–H), for (C–C, Si–C, Ge–C and Sn–C), and for (C–H, Si–H, Ge–H and Sn–H).

- When there are lone pairs on X, the sequence is generally (but not always)

$$D(A_2–X) < D(A_3–X) > D(A_4–X) > D(A_5–X)$$

This trend exists for bonds from Group 14 (Group IV) atoms (C, Si, Ge, Sn) to chlorine and to nitrogen, and for bonds from Group 15 (Group V) atoms (N, P, As, Sb) to fluorine and oxygen.

- The sequence of bond enthalpies,

$$D(F–F) < D(Cl–Cl) > D(Br–Br) > D(I–I)$$

reflects the last trend identified above and has been attributed to the strong repulsion between the non-bonding lone pairs of electrons on fluorine because of the short distance between the atoms.

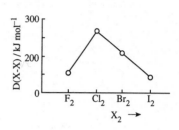

3.7 ΔH⚬ and reactivity

We started this chapter by contrasting the reactivity of NF_3 and NCl_3. Let us now examine their behaviour in respect of the following Hess cycle where X = F or Cl.

$$NX_3(g) \xrightarrow{-\Delta H_f(NX_3(g))} \tfrac{1}{2}N_2(g) + \tfrac{3}{2}X_2(g)$$

$$\Delta H^\circ = 3D(N–X) \qquad\qquad \Delta H^\circ = -\tfrac{3}{2}D(X\text{-}X)$$

$$N(g) + 3X(g) \xrightarrow{\Delta H^\circ = -\tfrac{1}{2}D(N\equiv N)} \tfrac{1}{2}N_2(g) + 3X(g)$$

In each case,

$$-\Delta H_f[NX_3(g)] = 3D(N-X) - \tfrac{1}{2}D(N\equiv N) - \tfrac{3}{2}D(X-X) \qquad (3.34)$$

Examination of Table 3.2 leads us to predict that the decomposition of NCl_3 will be much more exothermic than that of NF_3 since

- N–Cl bonds are weaker than N–F bonds, and
- the dissociation energy of F_2 is much less than that of Cl_2.

The contribution of $D(N\equiv N)$ is the same for both molecules. Thus the sign and magnitude of the reaction enthalpy change enables us to relate (by means of bond energies) observed differences in chemical reactivity — NCl_3 (eqn (3.2)) is explosive but NF_3 (eqn (3.1)) is not — *via* molecular properties of the molecules concerned. In this example, the relative weaknesses of the N–Cl bond compared with the N–F bond and that of the F–F bond relative to the Cl–Cl bond are the key factors.

3.8 Reactivity: is ΔH⚬ the only criterion?

We have seen that the relative behaviour of NF_3 and NCl_3 in respect of their decomposition to their component elements (N_2 and F_2 or Cl_2) reflects the enthalpy change of the reaction. For the fluoride (eqn (3.1)) the reaction is *endothermic* and $\Delta H^\circ = +125$ kJ mol^{-1}. In contrast for the chloride (eqn (3.2)) $\Delta H^\circ = -231$ kJ mol^{-1} and the reaction is strongly (explosively) *exothermic.* Developing these observations it is natural to ask: do all chemical reactions involve the evolution of heat?

Consider the addition of two different ionic solids, NaOH and NH$_4$Cl, separately to water so as to form a one molar solution. In the former case the dissolution is strongly *exo*thermic whereas in the latter case it is very *endo*thermic. Yet in both cases the solid very readily dissolves. It is clear that enthalpy changes are not the sole predictor of chemical change.

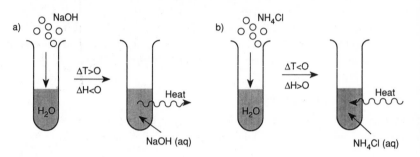

Fig. 3.5 Addition of solids to water: (a) exothermic reaction, (b) endothermic reaction.

The fact that chemical reactivity is not always associated with negative (exothermic) ΔH values is emphasized by a reaction which can proceed very vigorously, indeed violently, but which is endothermic, namely, the reaction between thionyl chloride and cobalt chloride:

$$6SOCl_2(\ell) + CoCl_2 \cdot 6H_2O(s)$$
$$\rightarrow CoCl_2(s) + 6SO_2(g) + 12HCl(g) \tag{3.35}$$

As with enthalpy changes, if entropy changes occur at SATP (298 K and 1 atm) we use the symbol ΔS^{O}.

The crucial quantity that we have so far neglected is the change in *entropy,* ΔS, associated with a chemical reaction.

The entropy of a substance may be thought as a measure of its *disorder.* Thus for any reaction, if

$$\Delta S > 0 \tag{3.36}$$

then it proceeds with an *increase* in disorder. Let us identify some processes in which this is the case.

*   A solid melts into a liquid. This is illustrated by the behaviour of ice at 10°C and 1 atmosphere pressure. In the solid the component atoms are ordered in a regular three-dimensional structure (e.g. Fig. 2.3) whereas in the liquid state they are free to move about and, at any moment, the instantaneous structure is comparatively disordered.

Fig. 3.6 illustrates how melting and boiling are accompanied by increases in disorder and hence of entropy.

*   A liquid vaporizes into a gas. This is illustrated by the behaviour of water at 110°C and 1 atmosphere pressure.
*   A molecule fragments into two or more smaller molecules. For example

$$N_2O_4(g) \rightarrow 2NO_2(g) \tag{3.37}$$

Two moles of gas have more disorder than one mole.

*   A solid decomposes into gaseous molecules. For example

$$NH_4Cl(s) \rightarrow NH_3(g) + HCl(g) \tag{3.38}$$

Gases are more disordered than solids.

• A reaction occurs which increases the number of moles of gaseous species. For example

$$C(s) + CO_2(g) \rightarrow 2CO(g) \qquad (3.39)$$

In any given reaction *both* the exothermicity ($\Delta H^o < 0$) and the increase in disorder ($\Delta S^o > 0$) are *driving forces* which encourage the reaction to occur.

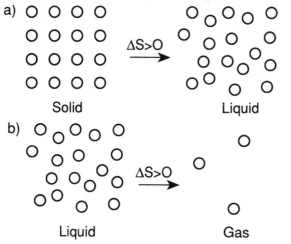

Fig. 3.6 (a) The solid to liquid transition, (b) the liquid to gas transition.

3.9 Entropy versus enthalpy

Consider two reactions both of which have favourable entropy changes and both of which are endothermic:

$$CaCO_3(s) \rightarrow CaO(s) + CO_2(g) \qquad (3.40)$$

$$\Delta H^o = +178 \text{ kJ mol}^{-1} \qquad \Delta S^o > 0$$

and

$$NH_4NO_3(s) + H_2O(\ell) \rightarrow NH_4^+(aq) + NO_3^-(aq) \qquad (3.41)$$

$$\Delta H^o = +26 \text{ kJ mol}^{-1} \qquad \Delta S^o > 0$$

The increase in entropy in the first case results from the added disorder due to the gaseous carbon dioxide whereas in the second case the solid turns into ions which are free to move randomly about the aqueous solution. In both these examples the two driving forces, ΔH^o and ΔS^o, are in conflict: the entropy change favours the products whereas the enthalpy change does not. Our chemical experience tells us that at SATP the first reaction does not occur whereas the second does. However if calcium carbonate is heated to 900°C the first reaction does then take place and CaO and CO_2 are formed.

The above observations suggest that there is a 'balance' between ΔH^o and ΔS^o and that the latter driving force becomes more potent at higher temperatures. To quantify these effects we introduce a new quantity which incorporates both the enthalpy and entropy terms. This is the change in the *Gibbs free energy*, ΔG^o, accompanying the reaction. It is defined by

Since ΔH^o is measured in kJ mol^{-1} it follows that ΔG^o possesses the same units. In contrast the units of ΔS^o are kJ mol^{-1}K^{-1}.

This calculation is only approximate since both ΔH^o and particularly ΔS^o do change somewhat with temperature. Nevertheless the answer is not unreasonably far from the experimentally observed value of 900°C = 1173 K.

The temperature sensitivity of entropies is to be expected given their relationship with disorder which is likely to increase with temperature as the molecules of a gas move about more vigorously than those of liquids and solids. Since the entropies of the reactants and products will therefore depend on temperature so will the change in entropy between them — and this change can be either an increase or decrease depending on the sensitivity of the entropy of the reacting and product species to the temperature.

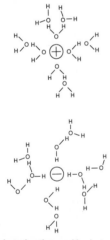

Hydrated cation and hydrated anion.

We have defined LE as the energy required to vaporize the solid to gaseous ions. The definition is sometimes taken to be the reverse process in which the LE's are all negative.

$$\Delta G^o = \Delta H^o - T\Delta S^o \qquad (3.42)$$

where T is the absolute temperature measured in Kelvin. If the above observations are generalized then it can be appreciated that *a decrease in free energy* ($\Delta G^o < 0$) *is the condition required for a reaction 'to go'*.

If we return to the pyrolysis of $CaCO_3$ we can now see why the decomposition temperature is near 900°C if we know that for the reaction of eqn (3.40) the standard entropy change is $\Delta S^o = 161$ J mol^{-1}K^{-1} and that ΔH^o is 178 kJ mol^{-1}. At room temperature,

$$\Delta G^o = \Delta H^o - T\Delta S^o = (178 - 298 \times 0.161) \text{ kJ mol}^{-1} > 0 \qquad (3.43)$$

and so $CaCO_3$ is stable. However, assuming ΔH^o and ΔS^o do not change much with temperature, then above a temperature

$$T > \frac{178}{0.161} = 1106 \text{ K} \qquad (3.44)$$

the free energy change becomes negative and decomposition of the carbonate is correctly predicted.

3.10 The dissolution of solids

Earlier in this chapter we saw that solid NH_4Cl dissolved endothermically in water whereas solid NaOH evolved heat upon dissolution. In this context it is helpful to define two enthalpies: the *lattice energy* (LE) of an ionic solid M^+X^- which is the energy required to vaporize one mole of the solid to form gaseous M^+ and X^- ions; and the *hydration energy* of M^+, $\Delta H^o_{hyd}(M^+)$, which is the enthalpy change when one mole of the pertinent gaseous ion is plunged into water at SATP. The hydration energy of X^-, $\Delta H^o_{hyd}(X^-)$, is similarly defined.

We can consider the following Hess cycle:

$$M^+X^-(s) \xrightarrow{\Delta H^o} M^+(aq)+X^-(aq) \qquad (3.45)$$

$$\Delta H^o = LE \qquad \Delta H^o = \Delta H^o_{hyd}(X^-)$$

$$M^+(g)+X^-(g) \xrightarrow{\Delta H^o=\Delta H^o_{hyd}(M^+)} M^+(aq)+X^-(g)$$

It follows that for the reaction of eqn (3.45)

$$\Delta H^o = LE + \Delta H^o_{hyd}(M^+) + \Delta H^o_{hyd}(X^-) \qquad (3.46)$$

where typically both the lattice and the hydration energies are large in magnitude whilst ΔH^o, as we have seen, is relatively small and can be either positive or negative in sign as illustrated earlier for NH_4Cl and NaOH. Table 3.3 shows the lattice energies for some of the alkali and alkaline earth metal halides.

The values in Table 3.3 lead to the following observations.
*   Doubly charged cations have higher lattice energies than singly charged cations.

• Small cations and anions, which can pack close to one another, as in LiF, form crystals with higher lattice energies than pairs of larger ions such as Cs^+ and I^-.

Table 3.3 Lattice energies for some alkali and alkaline earth metal halides. The structures are illustrated in Fig. 1.12.

Substance	Structure	LE / kJ mol^{-1}	Substance	Structure	LE / kJ mol^{-1}
LiF	NaCl	1025	CsI	CsCl	613
LiI	NaCl	760	MgF_2	TiO_2	2919
NaF	NaCl	907	$MgBr_2$	CdI_2	2402
NaCl	NaCl	777	MgI_2	CdI_2	2323
NaI	NaCl	706	CaF_2	CaF_2	2600
KCl	NaCl	706	$CaCl_2$	deformed TiO_2	2230
KI	NaCl	651	CaI_2	CaI_2	2075
CsF	NaCl	735	$SrCl_2$	CaF_2	2134
CsCl	CsCl	659	BaF_2	CaF_2	2323

Table 3.4 shows estimates of hydration enthalpies for various cations and anions. The following trends are apparent.
• The more highly charged the ion the greater the hydration energy.
• The smaller the gaseous ion the greater the hydration energy.
Both these factors reflect the extent to which the ion can attract water molecules to form hydrated ions. It is interesting to note that the factors affecting the hydration energies are similar to those controlling lattice energies. Thus it is not too surprising that a huge range of solubilities of ionic and other solids in water are observed, as discussed in Chapter 5.

$Li^+(aq)$ is larger than $Na^+(aq)$ as Li^+ holds more water molecules tightly around itself than does Na^+.

Table 3.4 Estimated values of single ion hydration enthalpies for various cations and anions.

Ion	ΔH^o_{hyd} / kJ mol^{-1}	Ion	ΔH^o_{hyd} / kJ mol^{-1}	Ion	ΔH^o_{hyd} / kJ mol^{-1}
H^+	−1095	Mg^{2+}	−1929	Fe^{3+}	−4393
Li^+	−517	Ca^{2+}	−1599	Ce^{3+}	−3381
Na^+	−407	Sr^{2+}	−1450	Ce^{4+}	−6513
K^+	−322	Ba^{2+}	−1303	F^-	−485
Rb^+	−297	Cr^{2+}	−1857	Cl^-	−342
Cs^+	−264	Cr^{3+}	−4418	Br^-	−315
Be^{2+}	−2496	Fe^{2+}	−1927	I^-	−274

4 Chemical kinetics

4.1 Introduction

In the previous chapter we saw that the direction of chemical reaction could be predicted using the concept of Gibbs free energy. Reactions proceed so there is a net decrease in the free energy of the reacting system: $\Delta G_{system} < 0$. Favourable, or 'thermodynamically possible' reactions are therefore those that occur with a decrease in enthalpy ($\Delta H_{system} < 0$) and/or an increase in entropy ($\Delta S_{system} > 0$). Examples of reactions which one would predict to be thermodynamically viable by these criteria include

$$Na(s) + H_2O(\ell) \rightarrow Na^+(aq) + OH^-(aq) + \tfrac{1}{2}H_2(g)$$
$$2H_2(g) + O_2(g) \rightarrow 2H_2O(g)$$
$$diamond(s) \rightarrow graphite(s)$$
$$aragonite(s) \rightarrow calcite(s) \tag{4.1}$$
$$2Al(s) + \tfrac{3}{2}O_2(g) \rightarrow Al_2O_3(s)$$
$$O_2(g) + haemoglobin(aq) \rightarrow CO_2(g) + H_2O(\ell) + ...$$

Only in the first case of eqn (4.1) does the reaction proceed spontaneously when the reactants are mixed. In the second case a spark is required to ignite the reaction, whereas in the other four cases the rate of reaction is extremely slow, at least in terms of the lifespan of a human being (which is about 10^9 s). In fact much of the world we live in is composed of a multiplicity of materials — sea-shells, wood, aluminium, apples, iron, ... — which are thermodynamically unstable with respect to chemical reaction with the atmosphere. Indeed, we ourselves are made largely of thermodynamically unstable components. It follows that chemists need to be able to understand the *rates* of chemical reactions as well as their thermodynamic (energetic) feasibility. This chapter is concerned with this essential topic.

4.2 The reaction pathway

Consider the reaction in which ferric ions oxidize iodide anions in aqueous solution:

$$2Fe^{3+}(aq) + 2I^-(aq) \rightarrow 2Fe^{2+}(aq) + I_2(aq) \tag{4.2}$$

Eqn (4.2) describes the overall chemical transformation. It does not describe the reaction mechanism — that is the sequence of individual chemical reactions by which the overall chemical transformation is brought about. In the case of interest the mechanism is thought to be two successive steps:

There are at least five polymorphic forms of $CaCO_3$ found in nature: aragonite and calcite are especially important. At SATP, calcite is less soluble than aragonite and hence is the thermodynamically more stable form. Nevertheless many carbonate sediments are composed to a large extent of aragonite. It is also the dominant form of $CaCO_3$ formed by organisms such as marine algae and is the material employed by the cuttlefish in its cuttlebone — an organ with a hollow porous structure which can be filled with water or gas to assist the creature either to sink or to float without having to expend energy by swimming continuously. The fascinating use of inorganic compounds as Nature's structural materials is described in *Biomineralisation*, S. Mann, J.Webb and R.J.P.Williams, VCH Publishers Inc. N.Y., 1989.

$$Fe^{3+}(aq) + I^-(aq) \rightarrow Fe^{2+}(aq) + I^{\bullet}(aq)$$
$$I^{\bullet}(aq) + I^{\bullet}(aq) \rightarrow I_2(aq)$$

(4.3)

In the first step, pairs of Fe^{3+} and I^- ions react to form $I^{\bullet}$ atoms and Fe^{2+} ions. Subsequently iodine atoms combine pairwise to form molecules. *The basic concept underlying chemical kinetics is that molecules must collide with each other in order to react*: no collision, no reaction. This partially explains why a sequence of steps, each of which requires collisions involving pairs of molecules, is required to effect the overall transformation. Quite simply the probability of a three-partner collision is much less than a two-partner collision since the chance of three species arriving simultaneously at a single point in space is considerably less than the arrival of two. The chance of a four-partner collision is even more remote. The reaction therefore follows the mechanism suggested above rather than the one implied by the stoichiometry of the reaction in eqn (4.2). The energy changes involved can be summarized in the reaction diagram (or profile) of Fig. 4.1.

> Radicals (see eqn (3.4)) are usually much more reactive than atoms or molecules where all the electrons are spin paired.

> Stoichiometry is the count of the number of reactant and product molecules involved in the reaction.

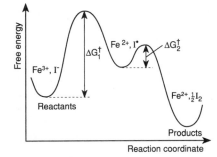

Fig. 4.1 A reaction profile.

Notice that the vertical-axis of the plot in Fig. 4.1 represents energy; the horizontal axis is loosely described as a *reaction coordinate* (the precise nature of this quantity need not be defined for most purposes). The multistep nature of the reaction is apparent and the reaction proceeds *via* $I^{\bullet}$ as an *intermediate* species. Notice that the (free) energy of the products $Fe^{2+}(aq)$ and $I_2(aq)$ lies below that of the reactants $Fe^{3+}(aq)$ and $I^-(aq)$ since the reaction is thermodynamically 'downhill'.

Fig. 4.1 shows that although the reaction occurs with a net decrease of free energy it requires a certain free energy, $\Delta G_1^{\dagger}$, to be given to the reactants before the reaction can proceed. This quantity of energy is termed the *free energy of activation* and the point of maximum energy on the reaction profile is known as the *transition state* for the reaction in question. Notice in the case depicted that the *activation energy*, $\Delta G_1^{\dagger}$, required to pass from the reactants to the intermediates

> An *intermediate* along a reaction path is a species that is at a local energy minimum but is not stable enough to exist for a long period of time and so be considered a product.

> The free energy of activation that is required for the reaction to take place comes from collisions of the reactants both with the solvent molecules and with each other.

$$Fe^{3+}(aq) + I^-(aq) \xrightarrow{\Delta G_1^{\dagger}} Fe^{2+}(aq) + I^{\bullet}(aq)$$

(4.4)

is larger than that needed to pass from the intermediates into the products:

$$I^{\bullet}(aq) + I^{\bullet}(aq) \xrightarrow{\Delta G_2^{\dagger}} I_2(aq)$$

(4.5)

The overall reaction can proceed no faster than the speed of the rate limiting step since that represents the slowest chemical step in the whole reaction sequence. This is the origin of the term 'rate limiting' for which the alternative 'rate determining' is also used.

The transition state which is of the highest absolute energy in any reaction mechanism controls the rate of reaction and the corresponding chemical step in the mechanism is known as the *rate limiting step*. In the example given, the rate limiting step is the initial oxidation of eqn (4.4).

4.3 Intermediates and transition states

This reaction is imaginatively labelled as of the 'S_N2' type by organic chemists: S = substitution, N = nucleophilic (a nucleophile is an electron rich species that 'likes' species that are electron poor), 2 = bimolecular (see later in main text).

In the previous section the terms *intermediate* and *transition state* were introduced. The difference between these two chemical species can be understood with reference to the two reaction profiles plotted in Fig. 4.2. In the first reaction profile the reactants pass directly into the products *via* a *transition state* as in the reaction

$$4\text{-}BrC_6H_4NO_2 + Cl^- \rightarrow 4\text{-}ClC_6H_4NO_2 + Br^- \qquad (4.6)$$

conducted in a mixed solvent of ethanol and dioxan.

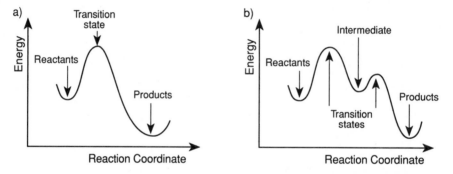

Fig. 4.2 Reaction profiles for (a) a one step reaction, and (b) a two step reaction *via* a metastable intermediate.

In the second reaction profile of Fig. 4.2 an intermediate is formed and transition states occur both before forming the intermediate from the reactants and before forming the product from the intermediate. An example of this type of reaction is the hydrolysis

This reaction is of the S_N1 type: 1 = unimolecular.

$$(CH_3)_3CCl \rightarrow (CH_3)_3C^+ + Cl^-$$
$$(CH_3)_3C^+ + H_2O \rightarrow (CH_3)_3COH + H^+ \qquad (4.7)$$

in aqueous ethanol solution. The carbonium ion, $(CH_3)_3C^+$, is the intermediate and the first step is rate limiting.

The key difference between an intermediate and a transition state is that the former has some degree of stability since it corresponds to a free energy minimum along the reaction profile whereas the latter is inherently unstable with respect to decomposition, being at a free energy maximum. Intermediates may therefore, if sufficiently long-lived, be isolated or otherwise can, at least in principle, be studied by some form of spectroscopy such as NMR (Chapter 2). Transition states of reactions in solution have only a fleeting existence and are not isolable, nor are they amenable to routine spectroscopic investigation.

4.4 Quantifying the rate of a reaction

Consider a solution of 2-chloro-2-methylpropane undergoing hydrolysis:

$$(CH_3)_3CCl + H_2O \rightarrow (CH_3)_3COH + H^+ + Cl^- \tag{4.8}$$

As the reaction proceeds the amount of $(CH_3)_3CCl$ in solution will diminish and its concentration — the number of moles of the haloalkane per unit volume — will decrease as shown schematically in Fig. 4.3.

$(CH_3)_3CCl \equiv Bu^tCl$, where Bu^t is the t-butyl group, $(CH_3)_3C-$.

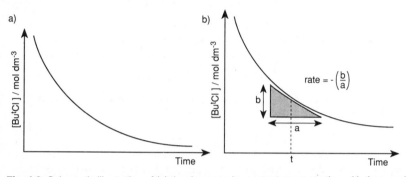

Fig. 4.3 Schematic illustration of (a) the decrease in reactant concentration with time, and (b) the calculation of the rate of loss of $(CH_3)_3CCl$ (denoted Bu^tCl) at time t.

The *rate of loss* of the reactant in Fig. 4.3 can be seen to change as the reaction continues. Initially the curve is steep and the rate of hydrolysis is greater than that observed later in the experiment where the curve is less steep. Mathematically the rate of loss of $(CH_3)_3CCl$ at any time may be found by drawing a tangent to the curve at the appropriate instant, t, to measure the gradient (rate of decrease of reactant) at that point (Fig. 4.3b). The quantity so evaluated, $-\left(\frac{b}{a}\right)$, has units of concentration divided by time or $mol\ dm^{-3}\ s^{-1}$. In terms of calculus the quantity is the first derivative of the concentration *versus* time plot:

$$\frac{d[(CH_3)_3CCl]}{dt} \bigg/ mol\ dm^{-3}\ s^{-1} = -\left(\frac{b}{a}\right) = \text{rate of loss of } (CH_3)_3CCl \tag{4.9}$$

$$= \text{rate of change of } [(CH_3)_3CCl]$$

In a typical experiment the solvent concentration, [solvent], will be tens of moles per dm^3 (pure water has $[H_2O] = 56$ M) but the Bu^tCl concentration will be at most around 10^{-2} M – 1 M. Therefore, since exactly one solvent molecule is lost for each reactant molecule the solvent concentration is only negligibly changed even when the reaction has run to completion. Mathematically,

$$\frac{d[\text{solvent}]}{dt} \cong 0$$

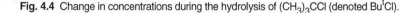

Fig. 4.4 Change in concentrations during the hydrolysis of $(CH_3)_3CCl$ (denoted Bu^tCl).

Such rates of reaction can be found for any chemical process provided concentration *versus* time plots are made for the reaction of interest — for example by using UV/visible spectroscopy to monitor the amount of reactant remaining in the solution. In the case of the 2-chloro-2-methylpropane hydrolysis, the only concentration which needs to be followed is that of the haloalkane since the water is the solvent and so is present in such vast excess that its concentration effectively does not change during the course of the reaction. Graphically, this may be illustrated as in Fig. 4.4.

The minus sign in eqn (4.10) can be appreciated if the concentration *versus* time plots are made for both species. The alcohol (Bu^tOH) concentration grows whilst that of the chloride (Bu^tCl) decreases.

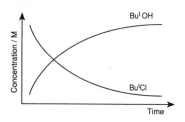

Notice also that, since one molecule of 2-methylpropanol is formed for each molecule of $(CH_3)_3CCl$ that reacts

$$\frac{d[(CH_3)_3COH]}{dt} = -\frac{d[(CH_3)_3CCl]}{dt} \tag{4.10}$$

Another chemical example which only requires the monitoring of a single reactant species is the decomposition of chlorate(I) in aqueous solution:

$$3ClO^-(aq) \rightarrow 2Cl^-(aq) + ClO_3^-(aq) \tag{4.11}$$

By measuring and then plotting $[ClO^-]$ as a function of time, the rate of reaction,

$$\frac{d[ClO^-]}{dt} \Big/ \text{mol dm}^{-3}\,\text{s}^{-1} \tag{4.12}$$

can be found using the method of tangents. In this example only one ClO_3^- anion is formed for every three chlorate(I) ions lost, so that

$$\left(\frac{1}{3}\right)\frac{d[ClO^-]}{dt} = -\frac{d[ClO_3^-]}{dt} \tag{4.13}$$

Likewise,

$$\left(\frac{1}{3}\right)\frac{d[ClO^-]}{dt} = -\left(\frac{1}{2}\right)\frac{d[Cl^-]}{dt} \tag{4.14}$$

If the rate of reaction in this example is found experimentally then again it is discovered that the rate decreases as the reaction continues. In fact the rate of reaction is discovered to be proportional to the square of the remaining ClO^- concentration,

$$\frac{d[ClO^-]}{dt} \propto [ClO^-]^2 \tag{4.15}$$

or

$$\frac{d[ClO^-]}{dt} = -k[ClO^-]^2 \tag{4.16}$$

In general if the rate law for the loss of a species A is given by

$$\frac{d[A]}{dt} = -k[A]^a$$

where k is a constant, the process is said to be a-th order in A. The units of k vary with the total order of the reaction.

where k is a positive constant. The parameter k in eqn (4.11) is known as the rate constant for the reaction; its units can be evaluated from eqn (4.16)

$$\text{units of } k = \frac{\text{units of reaction rate}}{\text{units of concentration squared}}$$

$$= \frac{\text{mol dm}^{-3}\,\text{s}^{-1}}{\left(\text{mol dm}^{-3}\right)^2} = \text{mol}^{-1}\,\text{dm}^3\,\text{s}^{-1} \tag{4.17}$$

Eqn (4.16) above is an example of a *rate law*. These are mathematical expressions that describe how the rate of reaction varies with the concentration(s) of the reactant(s). The particular example quoted is for a process which is second order in ClO^- since $[ClO^-]$ appears in the rate law as a squared term.

Rate laws relate reaction rates to the concentrations of the various chemical species involved in the process of interest. They are of great value to chemists since they can provide us with clear insights into the *reaction mechanism* by which the chemical reaction takes place. The following argument based on the ClO^- example illustrates this point.

We saw previously that a collision between reactants is required to bring about chemical reaction and that the simultaneous impact of more than two species was exceedingly unlikely. Moreover the fact that the reaction rate is *second order* in $[ClO^-]$ suggests that the transition state for the chlorate(I) decomposition involves simply the collision and reaction of *two* ClO^- anions. In contrast the overall reaction (eqn (4.11)) requires three anions to participate. All these observations can be rationalized in terms of the following reaction mechanism:

$$2ClO^-(aq) \xrightarrow{\text{slow}} ClO_2^-(aq) + Cl^-(aq)$$
$$ClO_2^-(aq) + ClO^-(aq) \xrightarrow{\text{fast}} ClO_3^-(aq) + Cl^-(aq)$$

$$(4.18)$$

The reaction profile is shown in Fig. 4.5. It may be summarized as follows:
- the species $ClO_2^-(aq)$ is an intermediate,
- the transition state is the combination of only two anions of ClO^- and is associated with an activation energy $\Delta G^\dagger$,
- the products $2Cl^-$ and ClO_3^- are of lower free energy than the reactants so that the overall transformation is thermodynamically viable.

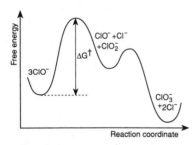

Fig. 4.5 Reaction profile for eqn (4.18).

It is important to appreciate that the experimentally observed second order rate law for the chlorate(I) decomposition is not consistent with the mechanism,

$$2ClO^-(aq) \xrightarrow{\text{fast}} ClO_2^-(aq) + Cl^-(aq)$$
$$ClO_2^-(aq) + ClO^-(aq) \xrightarrow{\text{slow}} ClO_3^-(aq) + Cl^-(aq)$$

$$(4.19)$$

This hypothetical process would have the reaction profile illustrated in Fig. 4.6 and it might be anticipated that this scheme would display third order kinetics,

$$\frac{d[ClO^-]}{dt} \propto [ClO^-]^3$$

$$(4.20)$$

since three molecules of ClO^- participate in the transition state, rather than the experimentally observed second order kinetics given by eqn (4.16).

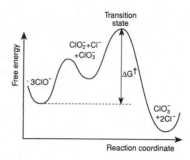

Fig. 4.6 Reaction profile for eqn (4.19).

If we return to the reaction in which aqueous iron(III) ions oxidize iodide anions (eqn (4.2)), then we might (correctly) expect that since the transition state of highest energy in the reaction profile of Fig. 4.1 involves simply the collision of one Fe^{3+} and one I^- anion, then

$$\frac{d[Fe^{3+}]}{dt} = \frac{d[I^-]}{dt} = -k[Fe^{3+}][I^-] \tag{4.21}$$

This rate law describes a process which is first order in both Fe^{3+} and I^- since the concentrations, $[Fe^{3+}]$ and $[I^-]$, appear with a unit exponent:

$$[Fe^{3+}] = [Fe^{3+}]^1; \ [I^-] = [I^-]^1 \tag{4.22}$$

A general rate law

Suppose we have a generalized reaction of stoichiometry,

$$aA + bB + cC + ... \rightarrow xX + yY + ... \tag{4.23}$$

Then the rates of appearance or disappearance of the various species are linked by the equation

$$Rate = R/(mol \ dm^{-3} \ s^{-1}) = -\frac{1}{a}\frac{d[A]}{dt}$$

$$= -\frac{1}{b}\frac{d[B]}{dt} = -\frac{1}{c}\frac{d[C]}{dt} = ... = \frac{1}{x}\frac{d[X]}{dt} = \frac{1}{y}\frac{d[Y]}{dt} = ... \tag{4.24}$$

The rate law may be expressed as

In general n_1 may or may not be equal to a, n_2 may or may not be equal to b etc.

$$R = -k[A]^{n_1}[B]^{n_2}[C]^{n_3} \tag{4.25}$$

where the process is order n_1 with respect to A, n_2 with respect to B, n_3 with respect to C and the overall reaction order, n, is defined as

$$n = n_1 + n_2 + n_3 + ... \tag{4.26}$$

Typical reaction orders are 0 (rate is independent of this species), 1, and 2. Fractional orders are also sometime found, as are negative values if the presence of a species *decreases* the reaction rate.

4.5 Reactions in solution

We have already considered the hydrolysis of 2-chloro-2-methylpropane under S_N1 conditions where the following mechanism operates

$$(CH_3)_3CCl \rightarrow (CH_3)_3C^+ + Cl^- \tag{4.27}$$

If the concentration of 2-chloro-2-methylpropane is recorded as a function of time then it can be shown that the hydrolysis obeys first order kinetics

$$R = -\frac{d[(CH_3)_3CCl]}{dt} = k[(CH_3)_3CCl] \qquad (4.28)$$

We have seen that for any system to react a certain energy — the activation energy — is required. In this example the source of this energy cannot be through collisions between the reactant molecules otherwise the process would be second order, rather than first order as is observed experimentally. Instead the energy required to overcome the activation barrier is supplied by occasional, highly energetic collisions from impacting solvent molecules. The latter molecules are not stationary but are moving about, constantly bumping into and off both one another and any solute molecules dissolved in the solvent. It is the more energetic collisions with the latter that provide the activation energy required to break the C–Cl bond necessary to ionize *t*-butyl chloride. The life and fate of a $(CH_3)_3CCl$ molecule which undergoes hydrolysis is illustrated in Fig. 4.7.

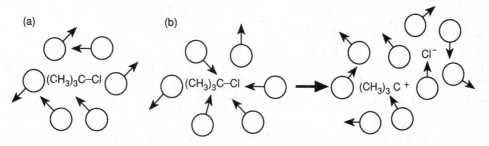

Fig. 4.7 $(CH_3)_3CCl$ colliding with solvent molecules (denoted by circles) leading to (a) no reaction and (b) hydrolysis. (a) No reaction takes place when solvent collisions with reactant are few and of low energy. (b) Reaction occurs when solvent collisions with reactant are many and of high energy.

Solvent influences are crucial in most liquid phase reactions. For example the S_N2 reaction

$$Cl^- + CH_3I \rightarrow I^- + CH_3Cl \qquad (4.29)$$

involves first the diffusion towards each other of the isolated reactants. Once the reactants are together the presence of surrounding solvent molecules makes their rapid separation difficult and the *encounter pair* is trapped inside this solvent cage for a period in which they can bounce on and off each other (maybe 100 – 1000 times) before one partner finds a route out of the 'cage' and the reactants separate without reaction as shown on the right-hand side of Fig. 4.8a.

On some occasions, however, collisions either within the encounter pair or with the surrounding solvent molecules are sufficiently energetic to provide the activation energy required for reaction. Nucleophilic displacement occurs as shown on the left-hand side of Fig. 4.8a for which the transition state is usually written as in Fig. 4.8b although the solvent, H_2O, molecules are often omitted. Nevertheless their role in the reaction, as we have seen, should certainly not be thought of as passive. Note also that

- the transition state involves one molecule of CH_3I and one chloride ion so that the reaction is first order in each and is second order overall,
- the dashed lines ||| in Fig. 4.8b represent partial bonding; the transition state is unstable with respect to decomposition *both* to Cl^- and to CH_3I (no reaction) or to I^- and CH_3Cl (nucleophilic substitution reaction).

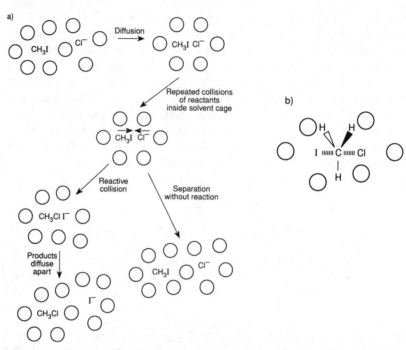

Fig. 4.8 (a) Schematic illustration of the eqn (4.29) reaction mechanism, (b) the transition state.

4.6 Reactions in the gas phase

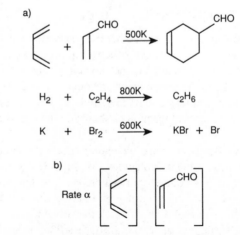

In the schematic representations of molecules in Fig. 4.9 each vertex stands for a carbon atom and the appropriate number of hydrogens (which are not explicitly shown in the figure) to give each carbon atom a total of four bonds.

Fig. 4.9 (a) Illustrative gas phase reactions. The first process is a Diels–Alder reaction, the rate equation for which is given in (b).

Illustrative gas phase reactions include the ones in Fig. 4.9. In this medium there is no solvent, so unless other inert gases (such as argon or nitrogen) are present the collisions which lead to reaction occur directly between the reactants so that *gas phase rate laws* are often first order in each reactant and overall second order (see Fig. 4.10).

Notice that the activation energy required for reaction must now come entirely from the collision between the two molecules participating in the reaction. Moreover when pairs of reactants come together and collide they either do so with sufficient energy to immediately react or else they collide, bounce apart, and then straightaway separate since there are now no solvent molecules to maintain an encounter pair. Fig. 4.10 shows the possibilities.

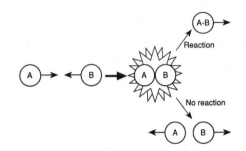

Fig. 4.10 Formation and fate of a gas phase collision.

Note that in most cases the collision involves the *direct* impact of the two molecules. In contrast, the reaction of potassium atoms, K, with molecular bromine, Br$_2$, is understood to proceed *via* a *harpoon mechanism*. At a separation of about 800 pm, before the two particles collide, the alkali metal atom ejects an electron — the 'harpoon' — towards the bromine molecule.

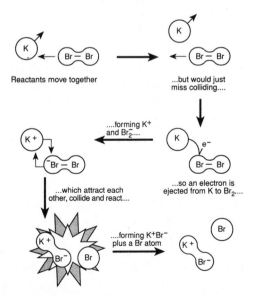

Fig. 4.11 K + Br$_2$ reaction and the harpoon mechanism.

The resulting K^+ and Br_2^- ions are then strongly attracted to each other by virtue of their new charges and are drawn close together: the 'harpoon is pulled in, and the final reaction occurs with the release of $Br^\bullet$ and the formation of the gaseous ion pair K^+Br^-. In this way impacting K and Br_2 species, which would otherwise have narrowly missed colliding, are forced to hit one another and to react (Fig. 4.11).

Rate laws other than the type cited above, which are first order in each reactant, can be observed in the gas phase. For example the reaction between nitric oxide and chlorine

$$2NO(g) + Cl_2(g) \rightarrow 2NOCl(g) \tag{4.30}$$

at 60°C shows an overall rate law of the form

$$R \propto -[NO]^2[Cl_2] \tag{4.31}$$

From our previous experience with reaction profiles we know this is telling us that in the transition state two molecules of NO and one of Cl_2 have combined. This is suggestive of the following two-step reaction mechanism:

$$2NO(g) \xrightarrow{\text{fast}} (NO)_2(g)$$
$$(NO)_2(g) + Cl_2(g) \xrightarrow{\text{slow}} 2NOCl(g) \tag{4.32}$$

The dimerization to form $(NO)_2$ is encouraged as NO is a radical species with an unpaired electron.

Another interesting gas phase reaction is the decomposition of azomethane:

$$CH_3N_2CH_3(g) \rightarrow N_2(g) + C_2H_6(g) \tag{4.33}$$

At low pressure of the gas this reaction displays second-order kinetics but at high pressures first-order kinetics are observed!

$$R \propto -[CH_3N_2CH_3]^n$$
$$n \sim 2 \text{ (low pressure)} \tag{4.34}$$
$$n \sim 1 \text{ (high pressure)}$$

The former limit can be readily understood if it is appreciated that the reactants need to collide in order that at least one of them becomes sufficiently energized in order to overcome the activation energy necessary for reaction:

Step 1 $\quad 2CH_3N_2CH_3(g) \xrightarrow{\text{slow}} CH_3N_2CH_3^*(g) + CH_3N_2CH_3(g)$

Step 2 $\quad CH_3N_2CH_3^*(g) \xrightarrow{\text{fast}} C_2H_6(g) + N_2(g)$

$$\tag{4.35}$$

where the superscript * denotes a highly energized molecule. The observed low pressure limit then simply reflects the fact that two molecules of azomethane occur in the transition state of the rate determing step (Step 1).

The high pressure behaviour of eqn (4.34) results because under these conditions the energized molecule is deactivated by a further collision

Step 3 $\quad CH_3N_2CH_3^*(g) + CH_3N_2CH_3(g) \rightarrow 2CH_3N_2CH_3(g) \tag{4.36}$

A simultaneous three-body collision between two NO molecules and one Cl_2 molecule is exceedingly unlikely.

In fact, because the $(NO)_2$ dimer formation is very fast compared to its subsequent rate of reaction the dimer has time to decompose back to NO and be reformed many times before the slow step takes place. The reaction is therefore more accurately drawn as

$$2NO(g) \rightleftharpoons (NO)_2(g)$$
$$(NO)_2(g) + Cl_2(g)$$
$$\xrightarrow{\hspace{2cm}} 2NOCl(g)$$

in which the first step is a *pre-equilibrium*.

Molecules possess different forms of energy — translational, rotational, vibrational and electronic energy. $CH_3N_2CH_3^*$ possesses much more of the first three types of energy than an ordinary $CH_3N_2CH_3$ molecule. It is not, however, usually electronically excited unless the reaction is a photochemical process and light is deliberately introduced to excite electrons within the molecule (see §2.7).

much more often than it reacts. The new rate determining step (Step 2) thus contains just a single molecule, $CH_3N_2CH_3{}^*$, as might have been expected on the basis of the discussion earlier in this chapter.

It is interesting to work out the net rate of formation of $CH_3N_2CH_3{}^*$. If k_1 and k_3 represent the rate constants for the second order reactions involved in Steps 1 and 3, respectively, whilst k_2 describes the first order reaction in Step 2 then,

$$\frac{d[CH_3N_2CH_3{}^*]}{dt} = k_1[CH_3N_2CH_3]^2 - k_2[CH_3N_2CH_3{}^*]$$

$$-k_3[CH_3N_2CH_3][CH_3N_2CH_3{}^*] \qquad (4.37)$$

However the energized molecule is a short-lived intermediate and so cannot build up in concentration during the reaction. In fact it rapidly reaches a steady state concentration such that

$$[CH_3N_2CH_3{}^*] \cong \text{constant} \qquad (4.38)$$

so that,

$$\frac{d[CH_3N_2CH_3{}^*]}{dt} = 0 \qquad (4.39)$$

It follows that

$$[CH_3N_2CH_3{}^*] = \frac{k_1[CH_3N_2CH_3]^2}{k_2 + k_3[CH_3N_2CH_3]} \qquad (4.40)$$

and

$$R = \frac{d[\text{products}]}{dt} = k_2[CH_3N_2CH_3{}^*] = \frac{k_1k_2[CH_3N_2CH_3]^2}{k_2 + k_3[CH_3N_2CH_3]} \qquad (4.41)$$

The recognition that for short-lived intermediates

$$\frac{d[\text{intermediate}]}{dt} \cong 0$$

is known as the steady state approximation. It is very useful, as here, in simplifying the analysis of kinetic problems involving complex mechanisms.

The scheme proposed here is known as the Lindemann mechanism.

The transition from second to first order kinetics can be understood since $[CH_3N_2CH_3]$ will increase as the pressure of azomethane is increased. Thus at low pressures

$$k_2 \gg k_3[CH_3N_2CH_3] \qquad (4.42)$$

so that

$$R = \frac{d[\text{products}]}{dt} \cong k_1[CH_3N_2CH_3]^2 \qquad (4.43)$$

and we see that second order kinetics are expected. This tells us again that the reaction rate is controlled by Step 1 and involves the collision of two azomethane molecules as the rate limiting step. However at sufficiently high pressures,

$$k_2 \ll k_3[CH_3N_2CH_3] \qquad (4.44)$$

and

$$R = \frac{d[\text{products}]}{dt} \cong \frac{k_1k_2[CH_3N_2CH_3]^2}{k_3[CH_3N_2CH_3]} = \left(\frac{k_1k_2}{k_3}\right)[CH_3N_2CH_3] \qquad (4.45)$$

so first order kinetics are seen.

4.7 The temperature dependence of reaction rates: collision theory and the Arrhenius equation

The number of collisions a single A molecule makes depends on v_R and [B].

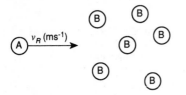

We have seen that collisions provide the necessary energy for surmounting the activation energy needed for a chemical reaction to occur. For the gas phase Diels–Alder reaction between buta-1,3-diene and 2-propenal (Fig. 4.9a)

$$\text{Activation energy} = +83 \text{ kJ mol}^{-1} \qquad (4.46)$$

Let A be a single butadiene molecule. The number of collisions it will make with propenal molecules, B, in a unit period of time depends on

- how fast, v_R (ms^{-1}), A is moving relative to the B molecules, and
- the number of B molecules per unit volume for it to hit.

Mathematically,

$$\text{Number of collisions} = v_R[\text{B}] \qquad (4.47)$$

It follows that the total number of collisions per unit volume in unit time will be given by

$$\text{Total number of collisions} = v_R[\text{A}][\text{B}] \qquad (4.48)$$

This result comes from the kinetic theory of gases.

The relative velocity, v_R, is known, for an ideal gas, to depend on the square root of the temperature according to:

$$v_R \propto \sqrt{T/\mu} \qquad (4.49)$$

where μ is the reduced mass of the molecular pair:

$$\frac{1}{\mu} = \frac{1}{m_A} + \frac{1}{m_B} \qquad (4.50)$$

The masses m_A and m_B are those of the two molecules, A and B. Thus the relative velocity is greater for lighter molecules at higher temperatures.

In order to calculate the reaction rate we need to know the number of adequately energetic collisions — that is the number with an energy greater than or equal to the activation energy, E_a. For an ideal gas the fraction, f, of collisions with energy greater than E_a is given by

It is conventional in writing the Arrhenius equation to denote the activation energy by E_a.

The exponential functions
$y = e^x = \exp(x)$ and
$y = e^{-x} = \exp(-x)$ are shown in the figures below.

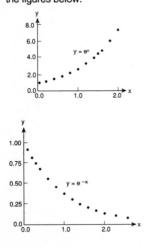

$$f = \exp\left(\frac{-E_a}{RT}\right) \qquad (4.51)$$

[Note we use R for the universal gas constant (eqn (1.8)) and R for rate.] It follows that

$$R = -\frac{d[\text{A}]}{dt} = -\frac{d[\text{B}]}{dt} \propto p\sqrt{\frac{T}{\mu}} \exp\left(\frac{-E_a}{RT}\right)[\text{A}][\text{B}] \qquad (4.52)$$

where we have now met all the quantities except, p, the steric factor; p is introduced to account for the fact that not all the sufficiently energetic collisions will have an appropriate orientation to lead to reaction. Thus for the Diels–Alder reaction above a rather specific orientation of the colliding species might be expected to be required for reaction. Thus most collisions could not lead to cyclization (Fig. 4.12).

Experimentally p is found to have a value of approximately 5×10^{-6} for the Fig. 4.12 system. This very low value indicates the severe stereochemical requirements for a successful collision. In contrast, for the reaction

$$\text{CH}_3{}^\bullet + \text{CH}_3{}^\bullet \rightarrow \text{C}_2\text{H}_6 \qquad (4.53)$$

where the orientation effects might be expected to be less critical, the value
of p is about 0.2 so that one in five adequately energetic collisions lead to the
formation of ethane.

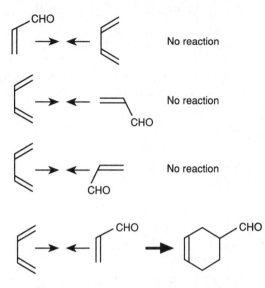

No reaction

No reaction

No reaction

Fig. 4.12 Reactive and non-reactive collisions.

The above theory is the so-called *collision theory* of reaction kinetics and
is directly applicable to gas phase reactions involving two species for which
the second order rate constant is predicted to be

> A simple reaction step involving two
> molecular species is referred to as
> 'bimolecular'.

$$k \propto p\sqrt{\frac{T}{\mu}}\exp\left(\frac{-E_a}{RT}\right) \qquad (4.54)$$

This expression shows that rate constants increase rapidly in size as the
temperature increases: synthetic chemists often perform reactions in refluxing
solvents to ensure that they occur at convenient temperatures. Since the $\sqrt{T}$
term in the above equation is much more weakly temperature dependent than
the exponential, a more widely used alternative to eqn (4.54) is the Arrhenius
equation which also describes reactions in solution:

> An Arrhenius plot for the thermal
> decomposition of HI(g) into H_2 and I_2.
> Notice that $\log_{10}$ has been used for the
> vertical axis (rather than $\log_e$) so that
> the slope is equal to
> $$\frac{-E_a}{2.303RT}$$
> This gives a value of
> $E_a = 185$ kJ mol^{-1}.

$$k = A\exp\left(\frac{-E_a}{RT}\right) \qquad (4.55)$$

$$\log_e k = \log_e A - \frac{E_a}{RT}$$

The second form of the Arrhenius equation tells us that if rate constants are
measured as a function of temperature then a plot of $\log_e k$ against T^{-1}
should be a straight line of slope $-E_a/R$. This protocol provides a means of
measuring activation energies.

The Arrhenius equation, eqn (4.55), tells us that rate constants are very
sensitive to temperature. Consider two values of k corresponding to two
temperatures T_1 and T_2 so that

$$\log_e\left(k_{T_1}\right) = \log_e(A) - \frac{E_a}{RT_1}$$

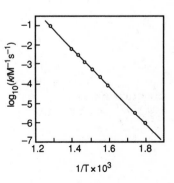

Here we assume that A does not change significantly for a small change in temperature.

$$\log_e\left(k_{T_2}\right) = \log_e\left(A\right) - \frac{E_a}{RT_2}$$

so $\quad \log_e\left(k_{T_2}/k_{T_1}\right) = \frac{E_a}{R}\left(\frac{1}{T_1} - \frac{1}{T_2}\right)$ (4.56)

$$= \frac{E_a}{R}\left(\frac{T_2 - T_1}{T_1 T_2}\right)$$

Consider the values $T_1 = 290$ K and $T_2 = 300$ K corresponding approximately to room temperature. For a typical reaction E_a is often about 50 kJ mol^{-1} which gives

$$\frac{k_{300}}{k_{290}} \cong 2 \qquad (4.57)$$

It follows that if a reaction has an activation energy close to 50 kJ mol^{-1} then a 10°C rise in temperature roughly doubles the reaction rate.

4.8 Catalysis

The reaction between $S_2O_8^{2-}$ and I^- anions in water may proceed *via* the following mechanism

$$S_2O_8^{2-}(aq) + I^-(aq) \xrightarrow{\text{slow}} SO_4^{-\bullet}(aq) + SO_4^{2-}(aq) + I^\bullet(aq)$$

$$SO_4^{-\bullet}(aq) + I^-(aq) \xrightarrow{\text{fast}} SO_4^{2-}(aq) + I^\bullet(aq) \qquad (4.58)$$

$$2I^\bullet(aq) \xrightarrow{\text{fast}} I_2(aq)$$

with the result that I^- is oxidized to molecular iodine at the same time as the $S_2O_8^{2-}$ dianion is reduced to sulfate(VI). That the first step is rate determining is indicated by the rate law:

$$R = -\frac{d[I^-]}{dt} = 2\frac{d[I_2]}{dt} \propto -[S_2O_8^{2-}][I^-] \qquad (4.59)$$

Fe^{2+} is a homogeneous catalyst in the $S_2O_8^{2-}$ / I^- reaction since the catalysed reaction occurs in a single phase.

If small amounts of either Fe^{2+} or Cu^{2+} cations are added to the solution the reaction is greatly speeded up although these ions undergo *no net change* and remain at the completion of the reaction in the *same quantity as originally added*. These ions behave as *catalysts* for the reaction in question. The mechanism by which they act can be understood as follows with respect to Fe^{2+}:

$$S_2O_8^{2-}(aq) + Fe^{2+}(aq) \rightarrow SO_4^{-\bullet}(aq) + SO_4^{2-}(aq) + Fe^{3+}(aq)$$

$$SO_4^{-\bullet}(aq) + Fe^{2+}(aq) \rightarrow SO_4^{2-}(aq) + Fe^{3+}(aq) \qquad (4.60)$$

$$2Fe^{3+}(aq) + 2I^-(aq) \rightarrow 2Fe^{2+}(aq) + I_2(aq)$$

Problem: Write down a mechanism for the catalytic action of Cu^{2+} on the $S_2O_8^{2-}$ / I^- system.

Summing these reactions gives

$$S_2O_8^{2-}(aq) + 2I^-(aq) \rightarrow 2SO_4^{2-}(aq) + I_2(aq) \qquad (4.61)$$

confirming that there is no net consumption of the catalyst Fe^{2+}.

Catalysts speed up reactions. They are widely employed by nature in many biochemical processes and by man in industrial or laboratory syntheses to facilitate transformations that would not otherwise occur at a fast enough rate to be useful. *However, catalysts can do nothing for reactions which are not thermodynamically viable.*

Another example of catalysis occurs when a mixture of nitrogen and hydrogen can be persuaded to form ammonia on an industrial scale through contact with a suitably treated iron surface at temperatures around 400 – 550°C and pressures of $10^2 - 10^3$ atm:

$$N_2(g) + 3H_2(g) \rightleftharpoons 2NH_3(g) \qquad (4.62)$$

This is an example of *heterogeneous catalysis* since it involves two separate phases: the solid iron and the gaseous mixture of N_2, H_2, and NH_3. The iron speeds up the reaction by assisting the dissociation of nitrogen into atoms. If this happens in the gas phase,

$$N_2(g) \rightarrow 2N(g) \qquad (4.63)$$

it requires a large amount of energy: 945 kJ mol^{-1}. However, the iron surface binds the N atoms thus modifying the reaction pathway, resulting in a *huge* reduction in the activation energy required as shown in Fig. 4.13.

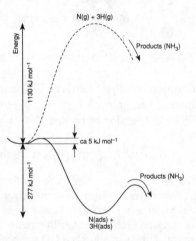

Fig. 4.13 The diagram shows the catalysis of the nitrogen plus hydrogen reaction and that both N_2 and H_2 dissociate to form adsorbed atoms on an iron surface. The symbol (ads) denotes an adsorbed species.

The nitrogen atoms are adsorbed on the iron surface at particular locations or sites. It has been found that rougher surfaces are more effective in promoting the N_2 dissociation, particularly if sites are available which can offer seven-coordination (seven nearest neighbours) to the adsorbed atom. Iron adopts a body centred cubic structure (Chapter 1) as shown in Fig. 4.14. Any iron surface may have different 'sections' through this structure. Thus the section ABFE is known as the (100) plane, the section ACGE as the (110) plane and the section ACH as the (111) plane. The surface atoms in these different planes have different numbers of nearest neighbours on the surface. Surface atoms are denoted C_i, where i denotes the number of nearest neighbours of an atom; the figure shows that only the (111) plane gives sites of seven-coordination to an adsorbed N. The indices (*hkl*) are known as the Miller indices of the planes.

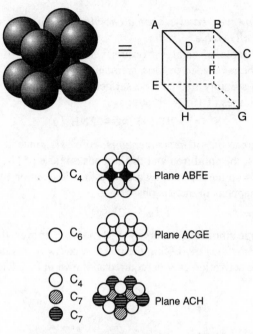

Fig. 4.14 The structure and surfaces of Fe(s).

The intelligent design of catalytic surfaces is a major area of industrial research. The following example illustrates the possibilities. Rhodium surfaces can catalyse the conversion of H_2 and CO into alkanes and alkenes. For example,

$$3H_2(g) + CO(g) \rightarrow CH_4(g) + H_2O(g) \tag{4.64}$$

However, the addition of atomic potassium to the surface increases the rate of CO conversion and encourages the formation of longer chain hydrocarbons. The potassium is said to act as a *promoter*. K atoms are thought to assist the reaction by each donating an electron to the the metal surface which in turn passes the added electron density to adsorbed CO molecules. This weakens the carbon–oxygen bond and encourages dissociation of carbon monoxide as shown in Fig. 4.15.

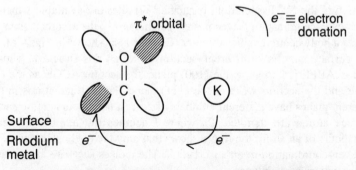

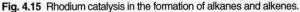

Fig. 4.15 Rhodium catalysis in the formation of alkanes and alkenes.

5 Chemical equilibria

5.1 Introduction: the equilibrium state

How far does a reaction go? What mixture of reactants and products will there be when the reaction has finished? What is the equilibrium state of this system? These are equivalent questions that are extremely important for all chemists. Sometimes we are interested in reactions such as the oxidation of methane (natural gas) in air

$$CH_4(g) + 2O_2(g) \xrightarrow{\Delta} CO_2(g) + 2H_2O(g) \qquad (5.1)$$

This reaction normally proceeds well, but when steam is present very little product is formed. Why should this be? Other important questions for our daily lives include: how much oxygen can be carried by blood? how much carbon dioxide can be dissolved in a fizzy drink? what is the maximum possible yield in an industrial process? Such questions are all governed by *chemical equilibria*.

If a system is in equilibrium, then the amount or concentration of all species present does not change with time. A number of features describe this state.

- In a true equilibrium, the system must be closed: material may not be added or removed — though a new equilibrium can be established after such interference. Open a bottle of fizzy drink and you disturb the equilibrium of carbon dioxide between the drink and the atmosphere — it goes flat!
- Variables such as temperature, T, pressure, P, and volume, V, will be constant. Again, changing these parameters can alter the position of equilibrium.
- Since we cannot stop two reactant molecules meeting and reacting, yet we require the amount of a species to remain constant, the process must be *reversible*. Any equilibrium is *dynamic*: reactants are changing into products at the same *rate* as products change into reactants.

If we add more reactants to an already equilibrated system, this leads to the formation of more products; if we add more products the amount of reactants present increases.

5.2 Recognizing the equilibrium state

Many important industrial processes finish when the equilibrium state has been established. It is important to be able to recognise when this has happened. For example carbon monoxide, CO, is produced by first reacting air with coke (carbon) to produce carbon dioxide, CO_2. The CO_2 then reacts

Δ indicates that the reaction requires heating to proceed.

Much chemistry occurs in *open* systems such as a blast furnace for iron production. For sufficiently slow flow rates of reactants through the reactor, the system approximates to a local equilibrium.

The dynamic nature of equilibria may be shown by the addition of isotopes to an equilibrated reaction mixture. For example, adding D_2 to an equilibrium involving H_2 as a reactant will lead to the appearance of D in the products.

The mixture of CO (25%), CO_2 (4%) and N_2 (70%), Ar *etc.* (1%) produced in this process is called producer gas — a useful industrial fuel.

with more coke, at 1100 K, in a very endothermic reaction that involves the net production of two moles of CO gas from every one mole of CO_2.

$$CO_2(g) + C(s) \rightleftharpoons 2CO(g) \tag{5.2}$$

$$\Delta H° = 171 \text{ kJ mol}^{-1}$$

$$\Delta S° = 176 \text{ J K}^{-1} \text{ mol}^{-1}$$

We could tell when no more CO is being produced by measuring the pressure in a constant volume, constant temperature reaction vessel.

A second gaseous equilibrium reaction involves nitrogen(IV) oxide dimerizing to form dinitrogen tetroxide. Here two molecules of reactant $NO_2(g)$ give one molecule of product $N_2O_4(g)$:

$$2NO_2(g) \rightleftharpoons N_2O_4(g) \tag{5.3}$$

brown   colourless

$$\Delta H° = -57.2 \text{ kJ mol}^{-1}$$

$$\Delta S° = -175.8 \text{ J K}^{-1} \text{ mol}^{-1}$$

In this case, the simplest method to determine when the equilibrium is established is to wait until the mixture is a constant colour. In general, we establish when an equilibrium has been attained by measuring any property that tells us that the concentration of any of the reactants or products is no longer changing. This may involve taking small samples of the reaction mixture at different times and analysing them until no further change is observed.

5.3 Predicting the equilibrium state: the equilibrium law

For any useful reaction it is essential to know what the equilibrium position is and also on what timescale the reaction will reach that equilibrium. For example, if in the earlier reaction for the production of CO, either only a few percent of the final gaseous mixture was CO or it took a week for the equilibrium to be reached, it would not be a viable industrial process. The timescale question can be resolved using the kinetic methods described in the previous chapter. We now consider the problem of the equilibrium position.

As we noted above, any equilibrium is in a 'dynamic state': it involves opposing 'forward' and 'backward' reactions that have no *net* effect on the concentrations of species present. At equilibrium, the *rate* of the forward step which takes reactants to products is equal to the backward step of products to reactants. Consider the reaction:

$$A + B \underset{k_r}{\overset{k_f}{\rightleftharpoons}} X \tag{5.4}$$

where k_f and k_r are respectively the rate constants for the forward and reverse reactions. The rate at which the reactants are used up is:

$$R(\text{forward}) = k_f[A][B] \tag{5.5}$$

and the rate for their production by the reverse reaction is

$$R(\text{reverse}) = k_r[X] \tag{5.6}$$

$NO_2(g)$ is a radical, with an unpaired electron on the N atom; this confers a bent shape with bond angle 134°. $N_2O_4(g)$ is a planar molecule with an N–N bond, and approximately the same ONO bond angles as $NO_2(g)$.

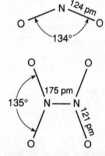

Stalactites (almost pure calcium carbonate formations which hang from the roofs of caves) grow over a period of hundreds or thousands of years from calcium hydrogen carbonate that is dissolved in rain water passing through limestone sediments. The aqueous calcium hydrogencarbonate (ground water) is in equilibrium with calcium carbonate (stalactite), carbon dioxide, and water.

$$Ca(HCO_3)_2(aq) \rightleftharpoons$$

$$CaCO_3(s) + CO_2(g) + H_2O(\ell)$$

Problem: What happens to stalactites if CO_2 or H_2O levels increase?

In writing eqn (5.4) we suppose that all of A, B, and X are in the same phase, *e.g.* in aqueous solution (aq), or in the gaseous state (g).

So, when an equilibrium is established:

$$k_f[\text{A}][\text{B}] = k_r[\text{X}] \tag{5.7}$$

Since, k_f and k_r are constants, we may write

$$K_c = \frac{k_f}{k_r} = \frac{[\text{X}]}{[\text{A}][\text{B}]} \tag{5.8}$$

> It is conventional to write *k* for *rate* constants and *K* for *equilibrium* constants.

where K_c is what is referred to as the *concentration equilibrium constant*. This equation is known as the *concentration equilibrium law* for the reaction of eqn (5.4).

For the general reaction

$$a\text{A}(\text{aq}) + b\text{B}(\text{aq}) + \ldots \rightleftharpoons x\text{X}(\text{aq}) + y\text{Y}(\text{aq}) + \ldots \tag{5.9}$$

it can be shown that

> Again it is assumed that A, B, ..., X, Y, ... are in one and the same phase.

$$K_c = \frac{[\text{X}]^x[\text{Y}]^y \ldots}{[\text{A}]^a[\text{B}]^b \ldots} \tag{5.10}$$

The units for each concentration in K_c are molar, M: moles of solute per dm^3 of solution. The units of K_c are then $\{\text{M}\}^{\Delta n}$, where $\Delta n = x + y - a - b$, the number of moles of solutes in the product side of the equation minus the number in the reactant side. For equal numbers of moles of reactants, $(a + b)$, and products, $(x + y)$, K_c has no units.

For gaseous reactions it is often convenient to use a different equilibrium constant, K_p, which has the same form as K_c, but uses *partial pressures* of each component rather than its concentration, so for the general case

> The ideal gas law (eqn (1.8)) may be applied to each component of a gaseous mixture:
> $$p_i V = n_i RT$$
> where p_i is the partial pressure of component *i* and n_i is its number of moles. Since the concentration of a species equals the number of moles divided by V (in dm^3) it follows that
> $$K_c = K_p(RT)^{a+b-x-y}$$
> $$= K_p(RT)^{-\Delta n}$$
> where $\Delta n = x + y - a - b$ and $R = 0.082056 \text{ dm}^3 \text{ atm K}^{-1}\text{mol}^{-1}$.

$$K_p = \frac{(p_\text{X})^x(p_\text{Y})^y \ldots}{(p_\text{A})^a(p_\text{B})^b \ldots} \tag{5.11}$$

where p_X is the partial pressure of X. When using K_p the units for each partial pressure are expressed in atmospheres, where 1 atm = 1.013×10^5 Pa. The units of K_p are then $\{\text{atm}\}^{\Delta n}$, where $\Delta n = x + y - a - b$.

> *Note:* $\Sigma_i \ p_i = P$

In using eqns (5.10) and (5.11) note the following.

- It is important to use *either* eqn (5.10) *or* (5.11), *not* a mixture of both (the exception to this arises for reactions in which soluble gases react with liquids as discussed in §5.5).
- The equilibrium constant depends on how the reaction equation has been written. For example, the equilibrium

$$\text{N}_2(\text{g}) + 3\text{H}_2(\text{g}) \rightleftharpoons 2\text{NH}_3 \tag{5.12}$$

has

> *Problem:* Write balanced equations for one mole of the first mentioned species, then write K_c and K_p for the gaseous equilibria, and show K_c has the units given in parentheses.
> (1) The cracking of ethane to give ethene and hydrogen {M}.
> (2) Synthesis of sulfur trioxide from sulfur dioxide and oxygen in the contact process {$\text{M}^{-1/2}$}.
> (3) The reaction of steam [$\text{H}_2\text{O(g)}$] and carbon monoxide to give hydrogen and carbon dioxide {no units}.
> (4) The gaseous reaction of hydrogen and oxygen to give water {$\text{M}^{-1/2}$}.

$$K_p = \frac{\{p(\text{NH}_3)\}^2}{\{p(\text{N}_2)\}\{p(\text{H}_2)\}^3} \tag{5.13}$$

but if we write the reaction so that only 1 mole of NH_3 is produced, the resulting K_p is the square root of that in eqn. 5.13. The unit of K_p change accordingly.

Uses of the equilibrium law

If K_c or K_p has been measured for a particular reaction at a given temperature we can use it to see 'how far a reaction will proceed'. If products predominate and the equilibrium lies 'far to the right', K can be very large, $K \gg 1$. Conversely, when very little product is produced its value is very much less than one, $K \ll 1$. For example, the reactions of the halogens with hydrogen to form halogen hydrides:

$$H_2(g) + X_2(g) \rightleftharpoons 2HX(g) \tag{5.14}$$

have the following values for K_p at 298 K:

X	F	Cl	Br	I
K_p	10^{47}	10^{17}	10^9	1

Thus this reaction is excellent for producing HF, but less so for HI. Note that in the above case K_p has no units since the reaction stoichiometry happens to be such that two moles of HX are formed for each pair of reactant molecules.

Because the equilibrium constants are unitless in this example, $K_p = K_c$

In this section we seek to illustrate how *equilibrium constants can be used to calculate the chemical composition of a reaction mixture that is at equilibrium*. To this end we suggest a general protocol that is useful in most calculations and then illustrate it with several examples. The protocol requires us to know how much of each reactant was present at the start of the reaction (t = 0) and also a value for K at the relevant temperature. The following is the recommended approach.

(1) Write out the chemical equation on one line.
(2) Write the concentration (or partial pressure) of each reactant present at time t = 0 beneath the chemical equation.
(3) On the next line write the equilibrium concentrations (or partial pressures) corresponding to t = ∞ in terms of a convenient parameter, x, which describes how much reactant has turned into product.
(4) Substitute these concentrations (or partial pressures) into the equation for K_c (or K_p) and determine x.
(5) Use the value of x to find the concentrations (or partial pressures) of all the species involved in the equilibrium.

Example 1

In aqueous solution iodide anions and molecular iodine reach an equilibrium in which triiodide is formed

$$I^-(aq) + I_2(aq) \rightleftharpoons I_3^-(aq) \tag{5.15}$$

for which $K_c = 759$ M^{-1} at 25°C. Let us calculate how much iodine remains in the solution as I_2 if 0.025 mol of potassium iodide and 0.05 mol of iodine are dissolved together in 1 dm^3 of water. We follow the protocol recommended above.

Step 1	$I^-(aq)$	$+$	$I_2(aq)$	$\rightleftharpoons$	$I_3^-(aq)$	
Step 2	0.025		0.05		0	t = 0
Step 3	0.025 − x		0.05 − x		x	t → ∞

Step 4
$$K_c = \frac{x}{(0.05-x)(0.025-x)} = 759$$

so
$$x = 759(0.05-x)(0.025-x)$$

or
$$1.3 \times 10^{-3} x = 1.25 \times 10^{-3} - 0.075x + x^2$$

$$0 = x^2 - 0.076x + 1.25 \times 10^{-3}$$

$$x = \frac{0.076 \pm \sqrt{(0.076)^2 - 5 \times 10^{-3}}}{2}$$

$$= \frac{0.076 \pm 0.028}{2}$$

$$= 0.024 \text{ moles}$$

The general solution of the quadratic equation
$$ax^2 + bx + c = 0$$
is
$$x = \frac{-b \pm \sqrt{b^2 - 4ac}}{2a}$$

We select the negative sign of the $\pm$ since otherwise $x > 0.05$ M which is physically impossible as it implies $[I^-] < 0$!

Returning to eqn (5.15)

$$I^-(aq) \quad + \quad I_2(aq) \quad \rightleftharpoons \quad I_3^-$$

Step 5   0.001   0.026   0.024   t→∞

Thus at equilibrium 96% of the iodine originally present is in the form of the triiodide ion (0.024 M); the free I_2 has a concentration of 0.026 M.

Example 2

Consider the formation of HI from its gaseous elements at 298 K starting with a partial pressure of j atm of each reactant in a 1 dm^3 flask. Let us ask what is the final composition of the equilibrium mixture.

The partial pressure of a gaseous species is (assuming the ideal gas law, eqn (1.8)) nRT/V atm where n is the number of moles of the species and V is the volume of the container.

Step 1 $H_2(g)$ $+$ $I_2(g)$ $\rightleftharpoons$ $2HI(g)$

Step 2 j / atm j / atm   0   t = 0

Step 3 $j - x$ $j - x$ $2x$ t → ∞

$$K_p = \frac{p_{HI}^2}{p_{H_2} p_{I_2}} = \frac{(2x)^2}{(j-x)^2} = 1$$

Step 4   or $(2x)^2 = (j-x)^2$

$$2x = \pm(j-x)$$

$$x = \tfrac{j}{3}$$

We note that the root
$$2x = -(j-x)$$
gives a negative partial pressure!

Returning to the equation

$$H_2(g) \quad + \quad I_2(g) \quad \rightleftharpoons \quad 2HI(g)$$

Step 5 $\frac{2j}{3}$ atm $\frac{2j}{3}$ atm $\frac{2j}{3}$ atm   t→∞

The partial pressure of each species is $\frac{2j}{3}$ atm. This example is very special since the number of each reactant molecule equals the number of product molecules. In general this is not the case and the calculation is a little more involved as illustrated in the next example. In addition, in this example K_p happens to equal unity at 298 K, which is most unusual.

Example 3

Let us calculate the equilibrium mixture corresponding to eqn (5.3) at a fixed *total* pressure of P atm when we start with pure N_2O_4. At 298 K

$$K_p = \frac{p_{N_2O_4}}{p_{NO_2}^2} = 7.1 \text{ atm}^{-1} \tag{5.16}$$

We begin by applying the first three steps of the now familiar protocol,

Step 1	$2NO_2(g)$	$\rightleftharpoons$	$N_2O_4(g)$	
Step 2	0		P	$t = 0$
Step 3	p_{NO_2}		$p_{N_2O_4}$	$t \to \infty$

We now need to find an expression for the two partial pressures. To do this we note that

$$p_{N_2O_4} + p_{NO_2} = P \tag{5.17}$$

and that the partial pressure of each species is proportional to the number of molecules of that species present. Moreover, for each N_2O_4 molecule which reacts *two* of NO_2 are formed. We therefore introduce the *degree of dissociation*, α, which describes the *fraction* of the N_2O_4 molecules which have reacted back to NO_2.
Thus

$$p_{N_2O_4} \propto P(1-\alpha) \quad \text{so} \quad p_{N_2O_4} = \frac{P(1-\alpha)}{A}$$
$$p_{NO_2} \propto P(2\alpha) \quad \text{so} \quad p_{NO_2} = \frac{P(2\alpha)}{A} \tag{5.18}$$

where A is a constant which can be evaluated using eqn (5.17)

$$P = \frac{P(1-\alpha)}{A} + \frac{P(2\alpha)}{A} \tag{5.19}$$

so that $A = (1+\alpha)$. Thus substituting eqn (5.18) into (5.16),

Step 4 $$K_p = \frac{[P(1-\alpha)/(1+\alpha)]}{[P(2\alpha)/(1+\alpha)]^2} = \frac{(1-\alpha^2)}{4P\alpha^2}$$

Upon rearranging this equation we find that

Step 5 $$\frac{1}{\alpha^2} = 4PK_p + 1$$

Problem: It is good practice to check that the results of any calculation are correct. Verify that the calculated partial presures in Example 3 satisfy eqn (5.16).

For the case where P = 1 atm and $K_p = 7.1 \text{ atm}^{-1}, \alpha = 0.18$. Thus

$$p_{NO_2} = 0.31 \text{ atm} \quad \text{and} \quad p_{N_2O_4} = 0.69 \text{ atm}$$

Factors influencing the magnitude of equilibrium constants

In Chapter 3 the standard Gibbs free energy change, ΔG^o, in a reaction was introduced and shown to be a balance of enthalpic and entropic factors:

Note the standard states in these equations.

$$\Delta G^o = \Delta H^o - T\Delta S^o \tag{3.42}$$

ΔG^o is related to K_p by the equation

$$\Delta G^\circ = -RT \ln K_p \qquad (5.20)$$

For solution phase reactions
$\Delta G^\circ = -RT \ln K_c$

Thus, the magnitude of ΔG° controls the size of K_p and hence whether the equilibrium favours reactants or products. This may be appreciated better if we insert values into eqn (5.20) and obtain, at 298 K

ΔG° / kJ mol^{-1}	K_p	
−50	6×10^8	products favoured
−25	2×10^4	
0	1	
+25	4×10^{-5}	
+50	2×10^{-9}	reactants favoured

In *enthalpy driven* reactions, the position of the equilibrium is controlled by the strength of the bonds made and broken. For example, as seen above, the equilibrium (eqn (5.14))

$$H_2(g) + F_2(g) \rightleftharpoons 2HF(g) \qquad (5.21)$$

lies very far to the right. The relevant bond energies are shown in the margin. Another example, where enthalpic effects dominate a very large unfavourable entropic effect, is provided by:

$$CaO(s) + CO_2(g) \rightleftharpoons CaCO_3(s) \qquad (5.22)$$

$$\Delta H^\circ = -178 \text{ kJ mol}^{-1} \qquad \Delta S^\circ = -161 \text{ J mol}^{-1}$$

which, at 298 K and atmospheric pressure, lies in favour of the products. Note, however, that because the entropy term in eqn (3.42) is weighted by the absolute temperature, at sufficiently high temperatures the entropy effect will dominate. Thus above 1100 K, $CaCO_3$ will decompose at one atmosphere pressure to form CaO and CO_2 (see §3.9).

In *entropy driven* reactions, the position of the equilibrium is dominated by the maximization of disorder. This is illustrated by the net production of one mole of gas in the formation of carbon monoxide from carbon dioxide and carbon (see eqn (5.2)) in a blast furnace at 1100 K. Despite the enthalpy requirement, the increase in the number of moles of gas in this reaction ensures that carbon monoxide forms at 1100 K.

An interesting example of a highly endothermic reaction occurs when barium hydroxide combines with ammonium thiocyanate; the two solids react to form barium thiocyanate, ammonia, and water and the temperature plummets — whereas in most reactions the reaction vessel heats up, this one becomes very cold. There is a substantial entropy increase.

$$2NH_4CNS(s) + Ba(OH)_2 \cdot H_2O(s) \rightleftharpoons$$

$$Ba(CNS)_2(s) + 3H_2O(\ell) + 2NH_3(g) \qquad (5.23)$$

The more reactive the halogen, the further to the right the eqn (5.14) equilibrium lies. This can be explained from consideration of bond dissociation enthalpies for the halogens and their hydrides:

Halogen	F	Cl	Br	I
$D(X_2)$ / kJ mol^{-1}	157	243	194	153
$D(HX)$ / kJ mol^{-1}	569	432	363	299

As discussed in §3.6, F_2 has an anomalously low bond dissociation enthalpy. For the hydrides the bond energies decrease down the group with increasing bond length.

Problem: Determine ΔH° for the formation of HI from H_2 and I_2 (Example 2 above).

The value slightly favours the products; however, entropy slightly favours the reactants, and the net result at 298 K is the K_c value of 1 which favours neither reactants nor products.

Does a catalyst change the position of the equilibrium?

The irrelevance of the presence of a *catalyst* for the *position* of the equilibrium can be derived quite easily. Let E_{cat} be the reduction in the activation energy due to the catalyst, and let the forward and backward activation energies, with no catalyst, be E_a^f and E_a^r respectively (Fig. 5.1). The Arrhenius equation for the rate constant in the presence of the catalyst is then:

$$k_f(\text{catalyst}) = A \exp\left[-\frac{E_a^f - E_{cat}}{RT}\right]$$

$$= A \exp\left[-\frac{E_a^f}{RT}\right] \exp\left[\frac{E_{cat}}{RT}\right]$$

(5.24)

Similarly

$$k_r(\text{catalyst}) = A \exp\left[-\frac{E_a^r - E_{cat}}{RT}\right]$$

$$= A \exp\left[-\frac{E_a^r}{RT}\right] \exp\left[\frac{E_{cat}}{RT}\right]$$

(5.25)

Thus,

$$K(\text{catalyst}) = \frac{k_f(\text{catalyst})}{k_r(\text{catalyst})} = \exp\left[-\frac{E_a^f}{RT}\right] \Bigg/ \exp\left[-\frac{E_a^r}{RT}\right]$$

$$= \frac{k_f(\text{no catalyst})}{k_r(\text{no catalyst})}$$

(5.26)

$$= K(\text{no catalyst})$$

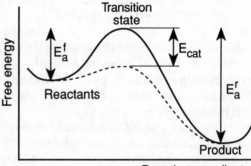

Fig. 5.1 Reaction energy profiles in the absence of a catalyst (solid line) and in the presence of a catalyst (broken line).

Le Chatelier's principle

Consider a chemical system at equilibrium. What happens when the pressure, the temperature, or the chemical contents of the reaction vessel are changed? Le Chatelier's principle summarizes the likely response: *if a system in equilibrium is exposed to a change, the equilibrium shifts to oppose that change.*

It is helpful to consider the effect of the most common changes that a chemist might impose on a system.

Changing the pressure: the pressure affects the position of an equilibrium if the number of moles of *gases* on the product side of the equation is not the same as on the reactant side. Increasing the pressure moves the reaction in the direction which produces fewer molecules, so the pressure increase is reduced (opposed). Pressure has little effect on reactions involving only liquids and solids.

Changing the temperature: the effect of this depends on whether the reaction is exothermic or endothermic. If *exothermic* the reaction gives out energy (as heat), so warming the reaction vessel increases the temperature and shifts the equilibrium in the direction that absorbs energy, which is towards the *reactants*. If the reaction is *endothermic*, then increasing the temperature favours the formation of the *products*.

Changing the volume: decreasing the volume of a reaction vessel for a gaseous reaction has the same effect as increasing the pressure.

Adding a catalyst: as discussed above, catalysts have *no* effect on the position of an equilibrium — they merely reduce the time it takes for the equilibrium to be established. Thus the equilibrium constant is unchanged.

Adding chemicals: additional reagents will affect the equilibrium position, though not the equilibrium constant. If reactants are added then some will react to form additional products, and conversely if products are added. Introducing new chemical species can lead to a new equilibrium.

Note the contrast between the equilibrium situation where Le Chatelier's principle applies, and the non-equilibrium situation of chemical kinetics where it does not. In the former, raising the temperature drives a reaction forward ($\Delta H^\circ > 0$) or backwards ($\Delta H^\circ < 0$). In the latter situation, raising the temperature usually *speeds up* the reaction regardless of whether ΔH° is positive or negative as heat provides the necessary activation energy to move molecules.

5.4 Writing equilibrium constants for gas–solid reactions

We have seen that for the gas phase reaction (Example 2, §5.3)

$$H_2(g) + I_2(g) \rightleftharpoons 2HI(g) \tag{5.27}$$

the equilibrium constant is defined by

$$K_p = \frac{p_{HI}^2}{p_{H_2}^2 p_{I_2}^2} \tag{5.28}$$

Likewise, for the liquid phase reaction

$$CH_3CH_2CO_2H(\ell) + CH_3CH_2OH(\ell) \rightleftharpoons$$
$$H_2O(\ell) + CH_3CH_2CO_2CH_2CH_3(\ell) \tag{5.29}$$

we have

$$K_c = \frac{[H_2O][CH_3CH_2CO_2CH_2CH_3]}{[CH_3CH_2CO_2H][CH_3CH_2OH]} \tag{5.30}$$

We consider next reactions involving gases *and* solids, for example

$$CO_2(g) + C(s) \rightleftharpoons 2CO(g)$$

Experiment shows in this case that

$$K_p = \frac{p_{CO}^2}{p_{CO_2}} \tag{5.31}$$

so that provided *some* carbon is present, the equilibrium constant is independent of the amount of graphite. Why this should be the case may be understood by considering the kinetics of the reaction which is likely to take place on the surface of the solid carbon which has an effective surface area, S. The forward reaction will require CO_2 to collide with the surface so the rate, R_f, may be written

$$R_f = k_f p_{CO_2} S \tag{5.32}$$

where k_f is a rate constant. Likewise the reverse reaction will require two carbon monoxide molecules to react on the carbon surface, so

$$R_r = k_r p_{CO}^2 S \tag{5.33}$$

where k_r is another rate constant. At equilibrium

$$k_f p_{CO_2} S = k_r p_{CO}^2 S \tag{5.34}$$

so that

$$\frac{k_f}{k_r} = \frac{p_{CO}^2}{p_{CO_2}} = K_p \tag{5.35}$$

Since the area of the solid carbon appears on *both* sides of eqn 5.34, changing the amount of carbon changes both the forward and the reverse rates by the same factor. Hence the independence of the final position of the equilibrium on the amount of carbon is understood.

In general, if we have a reaction

$$aA(g) + bB(g) + \ldots + eE(s) + fF(s) + \ldots \rightleftharpoons$$
$$mM(g) + nN(g) + \ldots + xX(s) + yY(s) + \ldots \tag{5.36}$$

then

$$K_p = \frac{p_M^m p_N^n \ldots}{p_A^a p_B^b \ldots} \tag{5.37}$$

and so K_p contains no terms in E, F, ..., X, Y, ...

Equally for reactions involving solids and aqueous solutions (or liquid mixtures),

$$aA(aq) + bB(aq) + \ldots + eE(s) + fF(s) + \ldots \rightleftharpoons$$
$$mM(aq) + nN(aq) + \ldots + xX(s) + yY(s) + \ldots \tag{5.38}$$

then

$$K_c = \frac{[M]^m [N]^n \ldots}{[A]^a [B]^b \ldots} \tag{5.39}$$

for similar reasons.

5.5 Special equilibria

In this section we shall look at some common situations where equilibria are particularly important. In some of these cases special vocabulary has been developed which makes them at first sight appear to be new subjects. However, there are no new *principles* involved, so they are no more nor less difficult than the examples we have considered so far.

Acid–base equilibria: competition for protons

The Lowry–Brønsted theory of acids and bases defines:

- an acid as a proton donor — it is a source of H^+ ions
- a base as a proton acceptor — it receives H^+ ions.

For example, when gaseous HCl is added to water the following equilibrium is established:

$$HCl(g) + H_2O(\ell) \; \rightleftharpoons \; Cl^-(aq) + H_3O^+(aq) \qquad (5.40)$$

The HCl(g) is acting as an acid by donating a proton to the base, H_2O, to form the hydronium ion H_3O^+(aq). Cl^- is called the *conjugate base* of hydrogen chloride; H_3O^+(aq) is the *conjugate acid* of water. Any acid–base reaction thus involves two conjugate acid–base pairs.

HCl(aq) is a *strong acid*, since it is highly dissociated in aqueous solution. Of all the known acids, only a few are strong: for example, HNO_3, H_2SO_4 (for the first proton only), HCl, HBr, $HClO_4$ (chloric (VII) acid), and HI. In contrast, ethanoic acid, CH_3COOH, and carbonic acid, H_2CO_3, are *weak acids* since they poorly dissociate in water.

Acid strength

In general, the strength of an acid, HA(aq), is given by the position of the equilibrium in its hydrolysis, with water acting as the base:

$$HA(aq) \; \rightleftharpoons \; A^-(aq) + H^+(aq) \qquad (5.41)$$

The stronger the acid, the further this equilibrium lies to the right; this is measured by K_c. The K_c for acids is labelled as K_a

$$K_a = \frac{[H^+(aq)][A^-(aq)]}{[HA_{aq}]} \qquad (5.42)$$

pH and pK_a

Since acids provide protons, the simplest method experimentally to calibrate the strengths of acidic aqueous solutions is to determine the concentration of hydronium ion. This concentration can vary by many powers of ten: the strongest acids can have $[H_3O^+(aq)]$ in the region of 10 M, whereas the strongest bases have $[H_3O^+(aq)] = 10^{-15}$ M. To cope with sixteen orders of magnitude (the ratio is 10 000 000 000 000 000 : 1) we use a *logarithmic scale*, and define

$$pH = -\log_{10}[H_3O^+(aq)] = -\log_{10}[H^+(aq)] \qquad (5.43)$$

The minus sign is used in the definition, since acid concentrations are usually less than 1 M, so we then work with positive numbers for pH. It should be noted that the *smaller* the pH value the *greater* the hydronium ion concentration in solution (Fig. 5.2).

The proticity of an acid gives the number of replaceable protons in its molecule. Thus ethanoic acid, CH_3COOH, is monoprotic because only the carboxylic hydrogen can be replaced. Sulfuric acid, H_2SO_4, is a diprotic acid, with two (relatively easily) removable protons per molecule of acid. In fact the second proton is less easily lost than the first: sulfuric acid is a strong

---

The aqueous proton in solution is associated with at least one water molecule making H_3O^+(aq), the so-called hydronium ion illustrated below. This is itself further hydrated. The symbols H^+(aq) and H_3O^+(aq) are both used for aqueous protons.

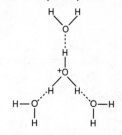

The expression for K_a can be deduced from the kinetics of acid dissociation and formation

$$HA(aq) \underset{k_r}{\overset{k_f}{\rightleftharpoons}} A^-(aq) + H^+(aq)$$

At equilibrium

$$k_f[HA] = k_r[A^-][H^+]$$

so

$$\frac{[A^-][H^+]}{[HA]} = \frac{k_f}{k_r} = K_a.$$

The decadic logarithm, $\log_{10}N$, of a number N is the power to which 10 must be raised so that it equals the number

$$10^{\log_{10}N} = N$$

Thus 10 000 000 000 000 000 is 10^{16} and its decadic logarithm is 16; 0.001 is 10^{-3} and its logarithm is –3. In summary,

N	$\log_{10}N$
100	2
50	1.7
10	1
5	0.7
1	0 †
0.1	–1
0.01	–2

† Recall that $10^0 = 1$.

acid for the first proton only. Hence 1.0 M sulfuric acid has a pH of –0.005: the second proton makes little additional contribution to the pH.

$$[H_3O^+]/M \quad 10 \quad 1 \quad 10^{-1} \quad 10^{-2} \quad \ldots \ldots \quad 10^{-7} \ldots \ldots \quad 10^{-12} \quad 10^{-13} \quad 10^{-14}$$

$$pH \qquad \quad -1 \quad 0 \quad 1 \quad 2 \quad \ldots \ldots \quad 7 \quad \ldots \ldots \quad 12 \quad 13 \quad 14$$

Fig. 5.2 Relationship between hydronium ion concentration and pH.

The pKa values of some acids are

H_2O_2	+11.7
H_3BO_3	+9.3
HOCl	+7.5
NH_4^+	+9.2
CH_3COOH	+4.8
$CH_3(CH_2)_2COOH$	+4.2
HCOOH	+3.7
H_2CO_3	+3.6
HF	+3.2
HSO_4^-	+1.9
HCl	–7.0
HBr	–9.5
HI	–10.0

The values for $HClO_4$, H_2SO_4, HNO_3 are *very* negative and difficult to measure accurately.

Problem: Check the pH of 1 M H_2SO_4 by assuming the first proton is fully dissociated, thus providing a 1M acid solution, and using $K_a = 1.2 \times 10^{-2}$ for the second proton.
Answer: –0.005 *not* –0.05 or –0.30.

K_a values also range over orders of magnitude, so it is also convenient to define pK_a by analogy with pH:

$$pK_a = -\log_{10}(K_a) \tag{5.44}$$

The weaker the acid, the higher its pK_a value. All K_a values refer to aqueous solution, unless specified otherwise. Eqn (5.42) may be rewritten

$$pH = pK_a + \log_{10}\frac{[A^-_{(aq)}]}{[HA_{aq}]} \tag{5.45}$$

and pK_a values may be determined by performing a pH *titration* of the acid with a very stong base (*e.g.* NaOH) and plotting pH versus added OH^-. Fig. 5.3 shows a typical curve with key features noted.

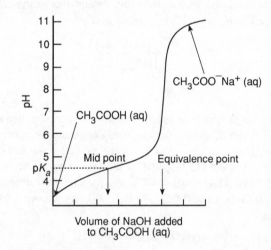

Fig. 5.3 pH curve for the titration of ethanoic acid with NaOH. At the equivalence point the number of moles of base added equals the amount of acid initially present. At the midpoint the concentrations of CH_3COOH and CH_3COO^- are equal so pH = pK_a (eqn (5.45)).

Acid dissociation constant of water, K_w

The equilibrium

$$2H_2O(\ell) \;\rightleftharpoons\; H_3O^+(aq) + OH^-(aq) \tag{5.46}$$

$$\Delta H^\circ = +56.8 \text{ kJ mol}^{-1}$$

is a very important one for all biological and many chemical systems. The equilibrium constant for the dissociation of water is therefore given a special symbol:

$$K_w = \left[H_3O^+(aq) \right]\left[OH^-(aq) \right] = 1.0 \times 10^{-14} \text{ M}^2 \text{ at 298 K} \qquad (5.47)$$

If no other acids or bases are present, then

$$[H_3O^+(aq)] = [OH^-(aq)] = 1.0 \times 10^{-7} \text{ so pH} = 7.0$$

This leads to the definition of a neutral solution at 298 K as one having pH = 7.0. However, this decreases with temperature since eqn (5.46) describes a reaction that is endothermic as shown in Table 5.1.

Table 5.1 The ionization constants of water as a function of temperature.

T/°C	$-\log_{10}K_w$	T/°C	$-\log_{10}K_w$
0	14.943	30	13.833
10	14.534	40	13.534
20	14.166	50	13.261
25	13.996	60	13.017

Base dissociation constant

Sometimes, instead of an acid dissociation constant, it is convenient to use a base dissociation constant, K_b, and the base A^-, in the equilibrium:

$$A^-(aq) + H_2O(\ell) \quad \rightleftharpoons \quad HA(aq) + OH^-(aq) \qquad (5.48)$$

where

$$K_b = \frac{[HA(aq)][OH^-(aq)]}{[A^-(aq)]} \qquad (5.49)$$

There is a simple connection between K_a and K_b for a conjugate acid/base pair HA(aq) and A^-(aq) that is obtained by multiplying the expressions for K_a and K_b

$$K_a K_b = K_w \qquad (5.50)$$

Or equivalently,

$$pK_a + pK_b = pK_w = 14 \text{ at 298 K} \qquad (5.51)$$

Buffer solutions

Control of pH is important, especially for biological systems. The human digestive system, for example, involves a range of pH values: pH = 6.4 – 6.8 in the mouth, pH = 1.6 – 1.8 in the stomach, and pH > 7 in the intestines. The pH of our blood is 7.35 – 7.45. Such restrictive pH ranges are achieved by the use of *buffers*. A buffer is a conjugate acid/base system whose equilibrium can shift to absorb or release protons, thus keeping the pH approximately constant. Any buffer contains significant concentrations of a weak acid and its conjugate base, usually from an added salt.

Referring back to the pK_a titration curve of Fig. 5.3 we can see that a weak acid will only be an effective buffer in the flat low pH region where

---

The absence of any [$H_2O(\ell)$] terms from eqn (5.47) may be understood from the kinetics of the reaction

$$2H_2O(\ell) \underset{k_r}{\overset{k_f}{\rightleftharpoons}} $$

$$H_3O^+(aq) + OH^-(aq)$$

At equilibrium

$$k_f[H_3O^+][OH^-] = k_r[H_2O]^2$$

so

$$[H_3O^+][OH^-] = \frac{k_f}{k_r}[H_2O]^2$$

$$= K_w$$

The [H_2O] term disappears into K_w for this case.

Problem: K_w increases with temperature, so the pH of water falls as it is warmed up. This is one reason for the increased corrosion found in hot water pipes compared with cold water ones. Use Table 5.1 to determine the pH of water at 0°C, 25°C, and 50°C.

The absence of [$H_2O(\ell)$] from eqn (5.49) is understandable in similar terms to those given above.

large volumes of base can be added for little pH change. The region where the system acts as a buffer is therefore close to the midpoint. In practice, as long as the ratio $[A^-]/[HA]$ is between, say, 0.1 and 10, then the system acts as a buffer. The log of the ratio of concentrations of the acid and its conjugate base is therefore small. To a reasonable approximation we can say that $[HA] \approx [HA]_0$, the amount of acid added to the solution and $[A^-] \approx [A^-]_0$. So eqn (5.45) becomes

$$pH = pK_a + \log_{10} \frac{[A^-(aq)]_0}{[HA(aq)]_0} \qquad (5.52)$$

which is known as the Henderson-Hasselbach equation for a buffer.

Because the last term in eqn (5.52) is generally small, a weak acid/conjugate base buffer acts at a pH in the range of its pK_a value. So ethanoic acid/sodium ethanoate buffers near pH = 4.77 — an important pH for many protein systems; ammonium ethanoate/ammonia buffers near pH = 9.3; and NaH_2PO_4 / Na_2HPO_4 (commonly referred to simply as phosphate buffer) acts in the neutral pH range. The pH of any buffer can be finely adjusted by altering the $[A^-]_0/[HA]_0$ ratio: more salt (M^+A^-) increases the pH (twice as much salt as acid gives a pH change of $\log_{10} 2 = 0.30$); more acid decreases the pH below the pK_a.

Acid–base indicators

An acid/base indicator is a weak acid, HA, whose conjugate base, A^-, is different in colour from the acid, so that when the pH of its solution changes the indicator signals the change. Usually such molecules are large aromatics whose π-electrons absorb visible energy with a high level of efficiency, so a very small concentration of HA has a very intense colour. Removing the proton from the indicator perturbs the π-electrons of the molecules so the colour of light that is absorbed changes. An indicator is often added to a solution to signify when its pH has changed. The colour of the indicator needs to be intense so that the number of protons required to cause the colour change is a very small fraction of those involved in the changes of the system we actually want to study.

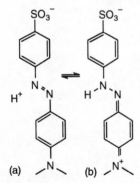

Methyl orange: (a) orange base and (b) red acid.

Table 5.2 Some common indicators, their useful pH range, and the colour change that occurs with increasing pH (decreasing proton concentration).

Indicator	pH range	colour change
methyl violet	0.0 – 1.6	yellow to blue
crystal violet	0.0 – 1.8	yellow to blue
methyl orange	3.2 – 4.4	red to yellow
methyl red	4.8 – 6.0	red to yellow
bromothymol blue	6.0 – 7.6	yellow to blue
3-nitrophenol	6.8 – 8.6	colourless to yellow
phenolphthalein	8.2 – 10.0	colourless to pink

The use of indicators may be understood by referring back to eqn (5.45). The indicator changes colour (described as the *end point* of the indicator) its solution passes from having more undissociated acid to having more free base present. So if we want to know when the pH value that equals the pK_a of the indicator is reached we just look for the colour change. It is of course important to select an indicator whose pK_a is at (or near) a point that is relevant for the system we are studying.

Since since the human eye needs about ten times as much HA as A^- (or vice versa) to be sure of the colour change a basic response requires
$$pH = pK_a + 1$$
and an acidic response requires:
$$pH = pK_a - 1$$
making the pH range for an indicator typically about 2 pH units. This is sufficiently accurate since we are usually trying to identify the midpoint (see Fig. 5.3) of a process where the pH change is very rapid.

Sparingly soluble salts

Sparingly soluble salts set up an equilibrium between the undissolved solid and the solvated ions in solution. For example, in the presence of water the silver halides, AgX for X=Cl, Br, or I, establish an equilibrium whose equilibrium constant is denoted K_{sp}, the solubility product:

$$AgX(s) \rightleftharpoons Ag^+(aq) + X^-(aq) \tag{5.53}$$

$$K_{sp} = \left[Ag^+(aq)\right]\left[X^-(aq)\right] \tag{5.54}$$

Values of K_{sp} for these systems at 298 K are 2.0×10^{-10} M^2 for AgCl, 7.7×10^{-13} M^2 for AgBr, and 1.5×10^{-16} M^2 for AgI. The steadily decreasing values reflect the increased level of covalent binding in the solid.

There is no mention of AgX(s) in eqn (5.54) since the reaction is of the solid / liquid type discussed earlier – see §5.4.

The common ion effect

When a second source of either the anion or cation is present we have what is known as the common ion effect which reduces the solubility of the salt. Silver chloride, for example, is much less soluble in 10^{-3} M NaCl than in water. Suppose s moles per dm^3 of AgCl dissolves in the aqueous NaCl solution, then $[Cl^-] = 10^{-3} + s$, and $[Ag^+] = s$. Now

$$K_{sp} = \left[Ag^+(aq)\right]\left[Cl^-(aq)\right] \tag{5.55}$$

so

$$(10^{-3} + s)(s) = 2.0 \times 10^{-10} \ M^2 \tag{5.56}$$

giving $s = 2.0 \times 10^{-7}$ M, rather than the value 1.4×10^{-5} M in the absence of the NaCl. The solubility of AgCl(s) is, however, largely unaffected by even large amounts of dissolved $NaNO_3$: there is no common ion in this case.

Problem: Write the equilibrium equation and an expression for the solubility product, with units, for each of the following sparingly soluble salts:
- calcium carbonate, $CaCO_3$,
- silver carbonate, Ag_2CO_3
- lead(II) iodide, PbI_2.

Problem: Use the values of K_{sp} for the silver halides given in the text to calculate their molar solubilities, s, at 298 K.

Problem: The common ion effect is even more marked for lead (II) chloride than for AgCl. Given that its $K_{sp} = [Pb^{2+}][Cl^-]^2 = 1.6 \times 10^{-5}$ M^3, determine the effect of 10^{-3} M and 0.1 M NaCl on the solubility of $PbCl_2$.

Predicting precipitation

Solubility involves a heterogeneous equilibrium between a solid and a liquid phase; the reverse process is precipitation. It is helpful to define the *ionic product*, which is defined in the same way as K_{sp} but applies to non-equilibrium situations. If the ionic product is greater than K_{sp} then too much salt is in solution and precipitation may occur. The ionic product is particularly useful for situations where a common ion is added to a solution. For example, if a given volume of 10^{-6} M silver nitrate is added in turn to the same volume of 10^{-6} M sodium chloride, bromide, and iodide solutions, should we expect some silver halide to be precipitated? In each case, remembering that the concentration of each solution is *halved* on mixing, we evaluate the ionic product to be $(5.0 \times 10^{-7}) \times (5.0 \times 10^{-7}) = 2.5 \times 10^{-13}$

M^2. Thus at 298 K, AgCl will not precipitate, AgBr just avoids doing so, and AgI is nearly all precipitated.

Reaction quotient

In using the ionic product we compare values of the equilibrium product for a given situation, K_{sp}, with the same formula in terms of products of concentrations, but calculated for a non-equilibrium situation. This concept may be extended to all systems for which an equilibrium constant may be measured and used to determine the direction of change expected for a non-equilibrium mixture.

We define the reaction quotient, Q, to have the same formula as the equilibrium constant, but to be evaluated in non-equilibrium situations. If

$$Q < K \tag{5.57}$$

then the reaction proceeds to increase Q. This involves the production of more products (which are on the top of the equilibrium constant equation) and less reactants. If

$$Q > K \tag{5.58}$$

then the converse occurs and the reaction proceeds towards the reactants.

Problem: Consider the iodine / iodide Example 1 in §5.3 above. If 500 cm^3 of a 0.25 M iodine solution, 250 cm^3 of a 0.25 M solution of I^-, and 250 cm^3 of a 0.25 M solution of I_3^- were combined, calculate Q, and by comparison with the equilibrium constant determine what, if any, reaction will take place.

Complex ion equilibria

Fig. 5.4 shows the solubility of silver chloride as a function of added chloride ion (in the form of, say, KCl). It can be seen that for low levels of added chloride the solubility decreases. This is due to the common ion effect discussed previously. In particular, the solubility, s, reflects how much Ag^+ is in the solution so that

$$s = [Ag^+] = \frac{K_{sp}}{[Cl^-]_{total}} \tag{5.59}$$

or

$$\log_{10} s = \log_{10} K_{sp} - \log_{10} [Cl^-]_{total} \tag{5.60}$$

where $[Cl^-]_{total}$ is the total chloride ion concentration, *i.e.* that from both the dissolution of AgCl and from the added KCl. Thus, since the scales of both axes in Fig. 5.4 are logarithmic, the *gradient* of the left hand part of the curve is (-1).

On the right hand side of the graph in Fig. 5.4, the solubility is seen to *rise* with increasing levels of chloride. The slope of the plot is close to $(+1)$. This increase is due to *complex ion formation*:

$$AgCl(s) + Cl^-(aq) \rightleftharpoons AgCl_2^-(aq) \tag{5.61}$$

The ion $AgCl_2^-$ has been characterized in the solid state and is known to be linear:

$$[Cl - Ag - Cl]^-$$

The complex ion formation may be described by an equilibrium constant

$$K_{ci} = \frac{[AgCl_2^-]}{[Cl^-]} \tag{5.62}$$

Since the equilibrium is of the solid / liquid type (see §5.4), there is no mention of solid AgCl in the expression for K_{ci}. In this region of the Fig. 5.4 curve, the solubility is essentially reflected by how much $AgCl_2^-$ is present in solution since the amount of Ag^+ is negligible — silver occurs as either solid AgCl or dissolved $AgCl_2^-$. Hence

$$s = [AgCl_2^-] = K_{ci}[Cl^-] \tag{5.63}$$

or

$$\log_{10} s = \log_{10} K_{ci} + \log_{10}[Cl^-]_{total} \tag{5.64}$$

The '+' on the right-hand side of eqn (5.64), in contrast to the '–' in eqn (5.60) explains the observed slope on the right-hand side of Fig. 5.4.

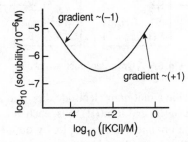

Fig. 5.4 AgCl solubility as a function of chloride ion concentration. Note the plot is log–log.

Redox equilibria: competition for electrons

Consider the reaction

$$Fe^{2+}(aq) + \tfrac{1}{2}Cl_2(g) \;\rightleftharpoons\; Fe^{3+}(aq) + Cl^-(aq) \tag{5.65}$$

in which iron(II) cations, Fe^{2+}, react with chlorine gas, Cl_2, to form iron(III) cations, Fe^{3+}, and chloride anions. The reaction involves the transfer of an electron with the result that

- Fe^{2+} loses an electron, and is *oxidized* to Fe^{3+}, and
- Cl_2 gains two electrons per molecule and is *reduced* to Cl^-.

The equilibrium constant for the reaction,

$$K = \frac{[Fe^{3+}][Cl^-]}{[Fe^{2+}]p_{Cl_2}^{1/2}} \tag{5.66}$$

lies in favour of the products. This fact might be established by conducting an experiment in which chlorine gas was passed through a solution of iron(II) salt. However, there is a much easier approach which involves the use of tabulated data determined from using *electrochemical cells*.

Fig. 5.5 shows a typical electrochemical cell. It comprises two half-cells linked by a *salt bridge*. The latter is simply a tube containing an aqueous potassium chloride solution (or a piece of soaked filter paper) to put the two half cells in electrical contact. Let us consider the two half-cells in turn.

On the left-hand side (as drawn) is a *standard hydrogen electrode*. It consists of a platinum electrode (coated in spongy platinum black) dipping into an

In the case of gas–liquid reactions, unusual equilibrium constants involving pressures and concentrations arise since the gas is soluble in the liquid (at least to some extent) so that for example

$$[Cl_2(aq)] \propto p_{Cl_2}$$

or $\quad [Cl_2(aq)] = \beta\, p_{Cl_2}$

Thus

$$K_c = \frac{[Fe^{3+}][Cl^-]}{[Fe^{2+}][Cl_2(aq)]^{1/2}}$$

$$K = \beta^{1/2} K_c$$

where β is a constant, describing the solubility of chlorine gas in water.

aqueous solution that is approximately 1 M in H^+ ions. Hydrogen gas — at a pressure of 1 atm — is bubbled through the solution and over the electrode.

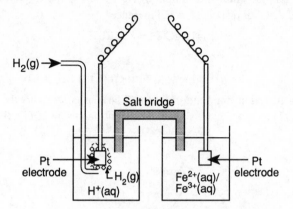

$H_2(g)$

Salt bridge

Pt electrode

$H_2(g)$

$H^+(aq)$

$Fe^{2+}(aq)/$
$Fe^{3+}(aq)$

Pt electrode

Fig. 5.5 A typical electrochemical cell.

On the right-hand side a platinum electrode dips into an aqueous solution which contains *both* iron(II) and iron(III) ions, each at a concentration close to 1 M. This might be obtained, for example, by dissolving $Fe(NO_3)_2$ and $Fe(NO_3)_3$ in water. Whilst the electrodes are unconnected, the cell is stable and no chemistry occurs. However, if the electrodes are short-circuited, that is directly connected to each other by a suitable conductor, then the following reaction occurs:

$$Fe^{3+}(aq) + \tfrac{1}{2} H_2(g) \rightleftharpoons Fe^{2+}(aq) + H^+(aq) \tag{5.67}$$

This is because electrons enter the left-hand electrode from H_2 gas, forming H^+ ions, then pass to the right-hand electrode where they reduce Fe^{3+} to Fe^{2+}.

The equilibrium constant

$$K\left(Fe^{3+}/Fe^{2+}\right) = \frac{[Fe^{2+}][H^+]}{[Fe^{3+}]p_{H_2}^{1/2}} \tag{5.68}$$

is large ($\log_{10}K = 13.0$) and so the equilibrium lies in favour of the products. The connecting wire simply provides a means of transferring charge (electrons) from one half cell to the other. Charge would flow until the reaction was brought to completion (by which we mean the equilibrium is established). However, if the cell is connected to a potentiometer, as shown in Fig. 5.6, then the point of connection on the contact wire can be adjusted until no charge flows through the ammeter (current measuring device). At this point, electron transfer between the two half-cells is stopped by an applied voltage which has magnitude

$$E^o = \left(\frac{\ell_2}{\ell_1} \times V\right) \text{ volts} \tag{5.69}$$

where ℓ_1 is the total length of the contact wire and ℓ_2 is the distance between the point of connection and the end of the contact wire which is

Note that in eqn (5.67) and all successive equations in this section the chemical reaction is written so that the reduced species gains *one* electron only and the oxidized species loses *one* electron.

In general the standard electrode potential of the redox couple $E^o(X^{(n+m)+}/X^{n+})$ can be found by forming a half-cell containing approximately 1 M $X^{n+}(aq)$ and 1 M $X^{(n+m)+}(aq)$ and using a potentiometer to measure the potential difference established when this half-cell is connected to a standard hydrogen electrode.

attached to the cell. V is the potential difference applied over the *whole* contact wire. The quantity E^0 is known as the *standard electrode potential* of the Fe^{3+}/Fe^{2+} couple and is a measure of how far the equilibrium of eqn (5.67) lies to the right. The value of $E^0(Fe^{3+}/Fe^{2+})$ is found to be +0.77 V.

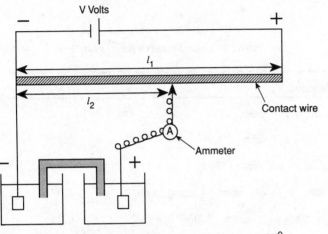

Fig. 5.6 Using a potentiometer for the measurement of E^0.

If the experiment is repeated replacing the Fe^{3+}/Fe^{2+} half-cell with a copper electrode dipping into a solution that is 1 M in $Cu^{2+}(aq)$ ions as shown in Fig. 5.7, then $E^0(Cu^{2+}/Cu)$ is found to be 0.34 V. This is a reflection of the extent to which the equilibrium

$$\tfrac{1}{2}Cu^{2+}(aq)+\tfrac{1}{2}H_2(g) \rightleftharpoons \tfrac{1}{2}Cu(s)+H^+(aq) \tag{5.70}$$

lies to the right. The equilibrium constant for eqn (5.70) is

$$K(Cu^{2+}/Cu)=\frac{[H^+]}{[Cu^{2+}]^{1/2}p_{H_2}^{1/2}} \tag{5.71}$$

In fact,

$$E^\circ(Cu^{2+}/Cu)=\tfrac{RT}{F}\ln K(Cu^{2+}/Cu) \tag{5.72}$$

$$E^\circ(Fe^{3+}/Fe^{2+})=\tfrac{RT}{F}\ln K(Fe^{3+}/Fe^{2+}) \tag{5.73}$$

where F is the Faraday constant and equal to the charge on one mole of electrons (approximately 96 500 C), R is the universal gas constant, and T is temperature in K.

Standard electrode potentials therefore tell us about equilibrium constants. In particular, if a half-cell involving two species M^{n+} and $M^{(n+m)+}$ is constructed, connected to a standard hydrogen electrode, and a potentiometer used to find $E^0(M^{(n+m)+}/M^{n+})$ then the equilibrium constant for the reaction

$$\tfrac{1}{m}M^{(n+m)+}+\tfrac{1}{2}H_2(g) \rightleftharpoons \tfrac{1}{m}M^{n+}(aq)+H^+(aq) \tag{5.74}$$

is given by

Note again that eqn (5.70) is written so as to involve the transfer of one electron.

Eqns (5.72) and (5.73) assume that the associated eqns (5.67) and (5.70) are written involving the transfer of only one electron. Otherwise the definition of K changes and the factor $\left(\tfrac{RT}{F}\right)$ is replaced by $\left(\tfrac{RT}{nF}\right)$, where n is the number of electrons transferred.

$$K\left(M^{(n+m)+}/M^{n+}\right) = \frac{\left[M^{n+}\right]^{1/m}\left[H^+\right]}{\left[M^{(n+m)+}\right]^{1/m}p_{H_2}^{1/2}}$$

$$= \exp\left(\frac{FE^\circ\left(M^{(n+m)+}/M^{n+}\right)}{RT}\right) \tag{5.75}$$

Table 5.3 Standard electrode potentials at 298 K.

Reaction	E°/V
$Li^+(aq) + e^- \rightleftharpoons Li(s)$	−3.04
$K^+(aq) + e^- \rightleftharpoons Na(s)$	−2.92
$Na^+(aq) + e^- \rightleftharpoons Na(s)$	−2.71
$\frac{1}{2}Mg^{2+}(aq) + e^- \rightleftharpoons \frac{1}{2}Mg(s)$	−2.37
$\frac{1}{3}Al^{3+}(aq) + e^- \rightleftharpoons \frac{1}{3}Al(s)$	−0.71
$\frac{1}{2}Zn^{2+}(aq) + e^- \rightleftharpoons \frac{1}{2}Zn(s)$	−0.76
$Cr^{3+}(aq) + e^- \rightleftharpoons Cr^{2+}(aq)$	−0.41
$\frac{1}{2}Fe^{2+}(aq) + e^- \rightleftharpoons \frac{1}{2}Fe(s)$	−0.41
$\frac{1}{2}Cd^{2+}(aq) + e^- \rightleftharpoons \frac{1}{2}Cd(s)$	−0.40
$H^+(aq) + e^- \rightleftharpoons \frac{1}{2}H_2(g)$	0.00
$\frac{1}{2}Cu^{2+}(aq) + e^- \rightleftharpoons \frac{1}{2}Cu(s)$	+0.34
$[Fe(CN)_6]^{3-}(aq) + e^- \rightleftharpoons [Fe(CN)_6]^{4-}(aq)$	+0.69
$Fe^{3+}(aq) + e^- \rightleftharpoons Fe^{2+}(aq)$	+0.77
$Ag^+(aq) + e^- \rightleftharpoons Ag(s)$	+0.80
$\frac{1}{2}Br_2(\ell) + e^- \rightleftharpoons Br^-(aq)$	+1.06
$\frac{1}{4}O_2(g) + H^+(aq) + e^- \rightleftharpoons \frac{1}{2}H_2O(\ell)$	+1.23
$\frac{1}{2}Cl_2(g) + e^- \rightleftharpoons Cl^-(aq)$	+1.36
$Ce^{4+}(aq) + e^- \rightleftharpoons Ce^{3+}(aq)$	+1.44
$Ag^{2+}(aq) + e^- \rightleftharpoons Ag^+(aq)$	+1.99
$\frac{1}{2}S_2O_8^{2-}(aq) + e^- \rightleftharpoons SO_4^{2-}(aq)$	+2.00

Notice that the species M^{n+} and $M^{(n+m)+}$ in eqn (5.74) need not necessarily be positively charged. For example, a half-cell can be formed by dipping a platinum electrode into a 1 M solution of chloride ions and bubbling over it chlorine gas at 1 atm pressure. The resulting value of $E^O(Cl_2/Cl^-)$ relates to the reaction

$$\tfrac{1}{2}Cl_2(g)+\tfrac{1}{2}H_2(g) \rightleftharpoons Cl^-(aq)+H^+(aq) \qquad (5.76)$$

for which

$$K\left(Cl_2/Cl^-\right)=\frac{[Cl^-][H^+]}{p_{Cl_2}^{1/2}p_{H_2}^{1/2}}$$

$$=\exp\left(\frac{FE^\circ\left(Cl_2/Cl^-\right)}{RT}\right) \qquad (5.77)$$

Other standard electrode potentials are given in Table 5.3.

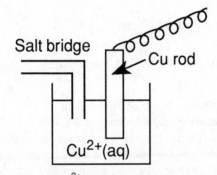

Fig. 5.7 Cu^{2+}/Cu electrochemical half cell.

Focusing on the general case identified in eqn (5.73), then we conclude that a positive value of the standard electrode potential means the equilibrium lies to the right. For a negative value it lies to the left. For example,

$$E^\circ\left(Li^+ / Li\right)=-3.04\ V \qquad (5.78)$$

So it follows that

$$Li^+(aq)+\tfrac{1}{2}H_2(g)\leftarrow Li(s)+H^+(aq) \qquad (5.79)$$

in accordance with chemical experience. From

$$E^\circ\left(Fe^{3+} / Fe^{2+}\right)=+0.77\ V \qquad (5.80)$$

it follows that

$$Fe^{3+}(aq)+\tfrac{1}{2}H_2(g)\rightarrow Fe^{2+}(aq)+H^+(aq)$$

as we have already seen (eqn (5.67)).

We started this section by considering the reaction (eqn (5.65))

$$Fe^{2+}(aq)+\tfrac{1}{2}Cl_2(g) \rightleftharpoons Fe^{3+}(aq)+Cl^-(aq)$$

for which

$$K = \frac{[Fe^{3+}][Cl^-]}{[Fe^{2+}]p_{Cl_2}^{1/2}}$$

$$= \left(\frac{[Cl^-][H^+]}{p_{H_2}^{1/2}p_{Cl_2}^{1/2}}\right) \div \left(\frac{[Fe^{2+}][H^+]}{[Fe^{3+}]p_{H_2}^{1/2}}\right) \tag{5.81}$$

$$= K\left(Cl_2/Cl^-\right) \div K\left(Fe^{3+}/Fe^{2+}\right)$$

$$= \exp\left\{\tfrac{F}{RT}\left[E^\circ\left(Cl_2/Cl^-\right) - E^\circ\left(Fe^{3+}/Fe^{2+}\right)\right]\right\}$$

Thus, a knowledge of $E^0(Cl_2/Cl^-)$ and $E^0(Fe^{3+}/Fe^{2+})$ enables us to predict the equilibrium constant for the reaction of interest. In this case

$$E^0(Cl_2/Cl^-) = 1.36 \text{ V and } E^0(Fe^{3+}/Fe^{2+}) = +0.77$$

so

$$K = \frac{[Fe^{3+}][Cl^-]}{[Fe^{2+}]p_{Cl_2}^{1/2}} \Big/ \left(\text{mol dm}^{-3}\text{atm}^{-1/2}\right)$$

$$= \exp\left\{\tfrac{F}{RT}[1.36 \text{ V} - 0.77 \text{ V}]\right\} = 9.4 \times 10^9 \Big/ \left(\text{mol dm}^{-3}\text{atm}^{-1/2}\right) \gg 1 \tag{5.82}$$

and the equilibrium lies in favour of Fe^{3+} and Cl^-.

Reflection on the above should convince the reader that Table 5.3 should enable us to predict the equilibrium constant for *any reaction composed of any two pairs of redox couples* listed in the Table. In particular

- redox couples with very negative standard electrode potentials contain good reducing *agents*, called reductants (which are themselves oxidized), *e.g.* Li is a strong reductant and is easily oxidized to Li^+
- redox couples with very positive standard electrode potentials contain good oxidizing *agents*, called oxidants (which are themselves reduced), *e.g.* Cl_2 is a strong oxidant and is easily reduced to Cl^-.

6 Taking it further

6.1 Introduction

In this chapter we shall briefly look at some particular examples that illustrate and extend the chemistry we have discussed in the previous chapters.

6.2 Michaelis–Menten kinetics

Without catalysts most of the chemistry of biological systems would not happen. Biological cells contain extraordinarily efficient catalysts known as enzymes. All enzymes are protein molecules and their catalytic activity is usually very specific. Proteases are one class of enzymes whose task is to *digest* proteins (*i.e.* break them into constituent parts). Chymotrypsin is a protease that cleaves proteins at points that leave a hydrophobic (water hating) amino acid residue at the carboxy terminus as illustrated in Fig. 6.1.

Not all biological catalysts are enzymes. The discovery of catalytic RNA has been one of the most exciting recent discoveries in biochemistry.

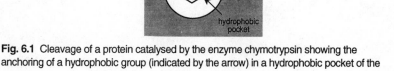

Fig. 6.1 Cleavage of a protein catalysed by the enzyme chymotrypsin showing the anchoring of a hydrophobic group (indicated by the arrow) in a hydrophobic pocket of the enzyme.

Enzyme reactions require the formation of a specific intermediate composed of the enzyme, E, and the reacting molecule (usually referred to as the *substrate*, S). This is followed by the reaction to form the product, P.

$$E + S \underset{k_r}{\overset{k_f}{\rightleftharpoons}} ES \xrightarrow{k_2} P + E \qquad (6.1)$$

The rate of product formation (eqn 4.24) is

$$R = k_2[ES]. \qquad (6.2)$$

If we define V_M to be the maximum possible rate (all the enzyme is bound in enzyme–substrate complexes) and $[E_o]$ to be the total enzyme concentration

$$V_M = k_2[E_o] \tag{6.3}$$

Problem: Derive eqn (6.5) using the
steady-state approximation (see
eqns(4.37)–(4.39)).

Since the intermediate *ES* is used up quickly we may use the steady state approximation (§4.6) so

$$\frac{d[ES]}{dt} = 0 \tag{6.4}$$

This results in

$$[E] = \frac{(k_r + k_2)[ES]}{k_f[S]} = \frac{(k_r + k_2)}{k_f} \frac{[ES]}{[S]} = K_M \frac{[ES]}{[S]} \tag{6.5}$$

where K_M is known as the Michaelis constant. So from eqn (6.5) the total enzyme concentration satisfies:

$$[E_o] = [E] + [ES] = \frac{K_M[ES]}{[S]} + [ES] = \left\{\frac{K_M}{[S]} + 1\right\}[ES] \tag{6.6}$$

which gives

$$[ES] = \frac{[E_o]}{\left\{\dfrac{K_M}{[S]} + 1\right\}} \tag{6.7}$$

Problem: Let $[S]_{1/2}$ be the substrate
concentration required to achieve half
the maximum rate. Show this equals
K_M.

eqns (6.2) and (6.7) then lead to

$$R = k_2[ES] = k_2[E_o]\left\{\frac{K_M}{[S]} + 1\right\}^{-1} = V_M\left\{\frac{K_M}{[S]} + 1\right\}^{-1} \tag{6.8}$$

V_M and K_M are constants that can be determined graphically and used to optimize the conditions for use of enzymes in many applications.

For example, the rate at which *N*-acetyl-L-valine methyl ester is hydrolysed by 10^{-5} M chymotrypsin (Fig. 6.1) varies with $[S]$ as

$R/(10^{-6}\ \text{mol dm}^{-3}\ \text{s}^{-1})$	0.40	0.50	0.58	0.70	1.00	1.30
$S/$ (M)	0.043	0.059	0.074	0.100	0.220	0.630

Problem: Determine what
concentration of substrate would be
required in order to get the
chymotrypin catalysis of *N*-acetyl-L-
valine methyl ester to proceed at 90%
of the maximum rate ($R_{90\%} = 1.41$).
Answer: $[S] = 1.125$ M.

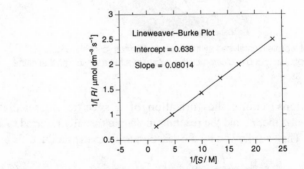

Fig. 6.2 Lineweaver–Burke plot of the chymotrypsin hydrolysis of *N*-acetyl-L-valine methyl ester.

If we rearrange eqn (6.8) to give

$$\frac{1}{R} = \frac{K_M}{[S]V_M} + \frac{1}{V_M} \tag{6.9}$$

and plot $1/R$ versus $1/[S]$, the intercept is $1/V_M$ and the gradient is K_M/V_M. From Fig. 6.2 we may then deduce that $V_M = 1.57$ μmol dm^{-3} s^{-1} and $K_M = 0.126$ M. Such a plot is a Lineweaver–Burke plot. Once we know the size of V_M we can decide whether it is feasible to speed up a particular reaction.

6.3 Electrolysis: reactions at the surfaces of electrodes

In electrolysis, a metal electrode connected to a voltage source acts as a source or sink of electrons — depending on the applied voltage — and is used to bring about chemical transformations. A current (flow of electrons) therefore flows through the electrode. The electrode acts as an anode if electrons pass from the solution phase to the electrode, for example

$$Fe^{2+}(aq) - e^- \rightarrow Fe^{3+}(aq) \qquad (6.10)$$

From §5.5 we see that since

$$E^0(Fe^{3+}/Fe^{2+}) = +0.77 \text{ V} \qquad (6.11)$$

the electrode voltage must be at least +0.77 V for this process to occur.

Conversely, an electrode functions as a *cathode* if electrons pass from it into a solution phase species, for example

$$\tfrac{1}{2}Ni^{2+}(aq) + e^- \rightarrow \tfrac{1}{2}Ni(s) \qquad (6.12)$$

Any cell must contain at least two electrodes. Thus in the *manufacture of* Cl_2 from brine (concentrated NaCl(aq)) the anode oxidizes chloride anions:

$$Cl^-(aq) - e^- \rightarrow \tfrac{1}{2}Cl_2(g) \qquad (6.13)$$

whilst the *cathode* reduces the solvent,

$$H_2O(\ell) + e^- \rightarrow \tfrac{1}{2}H_2(g) + OH^-(aq) \qquad (6.14)$$

so that the overall chemical change is

$$Cl^-(aq) + H_2O(\ell) \rightarrow \tfrac{1}{2}Cl_2(g) + \tfrac{1}{2}H_2(g) + OH^-(aq) \qquad (6.15)$$

The process thus produces two useful products; chlorine and sodium hydroxide. The annual world production of each by electrolysis approaches 40 million tons! The USA annual production (~ 10 million tons) requires 0.2 square miles of anode and 33 million megawatt hours of electric power!

Research chemists use *cyclic voltammetry* to study electrode reactions. In this technique the voltage of the electrode of interest is gradually increased (or decreased, depending on the system of interest), then the direction of the scan is reversed and the voltage steadily returned to its initial value. Throughout both the forward and reverse scans the current flowing is recorded as a function of voltage. When the voltage passes through a value which is sufficient for a new oxidation or reduction process to occur a peak is observed in the cyclic voltammogram (a plot of current against applied voltage).

Consider the reduction of C_{60} in CH_2Cl_2 solvent. The cyclic voltammogram is shown in Fig. 6.3. In this example the electrode voltage starts at 0.0 V and is then scanned in a negative sense to –2.0 V at which point the direction of the scan is reversed and the potential returned to 0.0 V. Four peaks are seen on the forward scan, corresponding to four reductions:

A current of one amp results from one coulomb of charge passing in one second. The charge on a single electron is 1.6×10^{-19} C. In the laboratory an electrolytic synthesis might use tens of milliamps whereas batteries of industrial mercury cells for NaCl electrolysis in the chlor-alkali industry have been built which pass up to 500,000 A .

The use of electrodes rather than chemical reducing or oxidizing agents to inject or remove electrons can be beneficial since the post-reaction separation of products is much easier, the lack of possibly toxic side-products formed from the otherwise added chemical agents is environmentally attractive, and a single electrolysis cell can be used for a wide variety of transformations.

Cl_2 is used to produce polyvinylchloride, chlorinated solvents, herbicides, methyl chloride (for lead alkyls), chlorofluorocarbons (CFCs) and for water treatment. The paper and pulp industry uses both Cl_2 and NaOH, and so often has on-site chlor-alkali plants to eliminate the environmental hazards of transporting these chemicals.

A brief exposure to 1 M NaOH can be used to skin peaches and potatoes prior to canning by the food industry.

The *direction* of current flow indicates whether the process is an oxidation or reduction. Note that positive currents — by convention — correspond to a flow of electrons in the opposite direction.

The voltammetry of fullerenes is discussed in papers by Dubois *et al, J.Am.Chem.Soc.* 113, (1991), 4364 and Jehoulet *et al.*, *J.Am.Chem.Soc.* 113, (1991), 5456.

Forward scan:
$$C_{60} \xrightarrow{+e^-(-0.44\,V)} C_{60}^- \xrightarrow{+e^-(-0.82\,V)} C_{60}^{2-}$$
$$\xrightarrow{+e^-(-1.25\,V)} C_{60}^{3-} \xrightarrow{+e^-(-1.72\,V)} C_{60}^{4-}$$
(6.16)

On the reverse scan the corresponding oxidations are also seen as current peaks but these are now inverted since the current flow is in the opposite direction from that in the forward scan:

Reverse scan:
$$C_{60}^{4-} \xrightarrow{-e^-(-1.72\,V)} C_{60}^{3-} \xrightarrow{-e^-(-1.25\,V)} C_{60}^{2-}$$
$$\xrightarrow{-e^-(-0.82\,V)} C_{60}^- \xrightarrow{-e^-(-0.44\,V)} C_{60}$$
(6.17)

In other organic solvents cyclic voltammetry reveals the formation of C_{60}^{5-} and C_{60}^{6-} anions.

In Fig. 6.3 for experimental convenience the voltage is measured against the Hg_2Cl_2/Cl^- redox couple rather than the H^+/H_2; the values would be approximately 0.24 V less negative on the hydrogen scale.

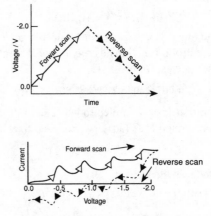

Fig. 6.3 Cyclic voltammogram for C_{60}.

6.4 Scanning tunnelling microscopy: looking at atoms and orbitals

In Chapter 1 the use of atomic force microscopy to observe atoms at the surface of graphite was described. A closely related technique is scanning tunnelling microscopy (STM). The basis of this experiment is as follows. A sharp metallic tip is brought to within less than 1 nm of a conducting surface and a small voltage is applied between the tip and the sample. A current flows and this is highly sensitive to the separation of the tip and the sample. Provided the separation is carefully maintained, atomically resolved images of the surface may be obtained as the tip is scanned over a selected area of the sample. STM is applicable to solids both in air and under liquids.

STM also provides a method for visualising *mos* (§2.5), as can be illustrated with reference to studies on elemental silicon. Silicon is isostructural with diamond (see Fig. 1.10) so that it might be thought that the solid surface would look as depicted in Fig. 6.4a. However this arrangement has two dangling (one electron) bonds per Si atom, so that it is energetically favourable for the surface to reconstruct forming the structure shown in Fig. 6.4b in which pairs of Si atoms move together and use one of the dangling bonds from each atom to form a σ bond. The remaining dangling bonds are approximately perpendicular to the silicon surface and can combine — as described in Chapter 2 — to form π and π^* orbitals. The energy level diagram of Fig. 6.5a. shows that the electrons reside in the π-orbital (the *highest occupied molecular orbital, homo*) with paired spins, whilst the π^* orbital (the *lowest unoccupied molecular orbital, lumo*) is unoccupied.

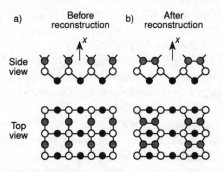

Fig. 6.4 (a) Si surface if it adopted a diamond structure, the dangling bonds are illustrated. (b) Si surface.

In the STM experiment, electrons leave the surface and and enter the tip if the voltage of the latter is positively biased relative to the sample. On the other hand, if the polarity is reversed (negative bias) then electrons must enter the silicon from the tip. In the former case they must come *from* the *homo* whereas in the latter situation they must transfer *to* the *lumo*. Thus the current pattern 'seen' by the STM in each case as the tip is moved over the surface should reflect the electron density in the two orbitals (given by the square of the wave function).

Schematic experimental results are shown are shown in Fig 6.5b in which the area depicted contains 4×6 $(Si)_2$ pairs. That twice as many 'spots' of high current are seen under negative bias as under positive bias provides very strong direct experimental evidence for the basis of the molecular orbital ideas developed in Chapter 2. In the silicon study the STM effectively visualizes the *mos*.

For a fuller discussion of the work described in this section see P. Avouros, *J.Phys.Chem.*, 94, (1990), 2246.

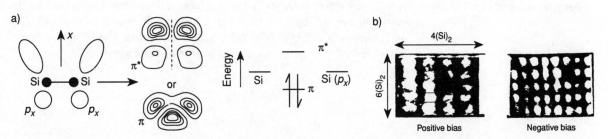

Fig. 6.5 (a) Energy level diagram and electron density for $(Si)_2$ p orbitals on the reconstructed silicon surface, (b) experimental data.

6.5 Drug design

The design of therapeutically active drugs is usually seen as the preserve of the pharmacologist or biochemist or the medicinal chemist. However, many of the most important aspects of this process follow directly from the physical chemistry foundation we have been developing in this book. To be effective a drug must

-     be able to be transported to its destination (its receptor)
-     not induce adverse side effects by reacting in the wrong part of the body
-     fit into the *active site*
-     perform the correct chemistry once delivered.

> The chymotrypsin hydrophobic pocket illustrated in Fig. 6.1 is an example of an active site.

The transport of a drug is dependent upon its shape and also whether it is sufficiently soluble in aqueous systems (such as the blood) and hydrophobic (water hating) systems (such as the membranes around cells through which many drugs have to pass). It is the kinetic and thermodynamic properties of a drug that determine whether it will be stable until it reaches its receptor. Sometimes a key geometric feature of a drug prevents it being metabolized in the wrong place because it then cannot fit into the 'wrong' receptor. For example, adding $-C{\equiv}CH$ to progestagens as illustrated in Fig. 6.6a reduces the extent to which they are metabolized in the liver.

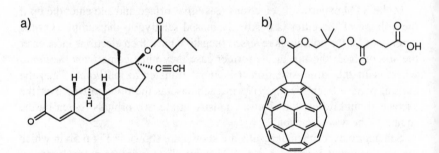

Fig. 6.6 (a) A progestagen, levonorgestrel butanoate, (b) a water soluble substituted–fullerene.

> Some of this fullerene research is described by S. Yamago *et al.* in *Chemistry and Biology*, **1995**, *2*, 385.

Another example is provided by fullerenes, to which we have referred throughout this book. Recently attempts have been made to develop fullerenes as drugs, either for their own intrinsic biological activity or for their capacity to act as cages or carriers for other drugs. The first challenge was to make them sufficiently water soluble. The molecule illustrated in Fig. 6.6b achieves this by the addition of a hydrophilic (water loving) 'tail'. The next question to ask is whether such a molecule is toxic and what side effects it may have. These questions are just beginning to be answered.

Further reading

Chemical Bonding; Winter, M. J., Oxford University Press: Oxford, 1994.

Chemistry and Chemical Reactivity; Kotz, J. C.; Purcell, K. F. Saunders, College Publishing: Orlando, 1991.

Chemistry: Principles and Applications; Atkins, P. W.; Clugston, M. J.; Frazer, M. J.; Jones, R. A. Y., Longman: London, 1988.

CRC Handbook of Chemistry and Physics; CRC Press: Cleveland, 1995.

The Elements of Physical Chemistry; Atkins, P. W., Oxford University Press: Oxford, 1992.

Essentials of Inorganic Chemistry; Mingos, D. M. P., Oxford University Press: Oxford, 1995.

Foundations of Organic Chemistry; Hornby, G. M.; Peach, J. M., Oxford University Press: Oxford, 1993.

Modern Physical Chemistry; Liptrot, G. F.; Thompson, J. J.; Walker, G. R., Bell and Hyman Limited: London, 1982.

Physical Chemistry (Fifth edition); Atkins, P. W., Oxford University Press: Oxford, 1994.

Quantities, Units and Symbols in Physical Chemistry (I.U.P.A.C.); Mills, I.; Cvitas, T.; Homann, K.; Kallay, N.; Kuchitsu, K., Blackwell: Oxford, 1993.

Revised Nuffield Advanced Science Book of Data; Ellis, H. (Ed.), Longman: London, 1984.

Index